The Essential Guide to
Hobby Farming
2nd Edition

Carol Ekarius with Leslie J. Wyatt

i-5
PRESS

The Essential Guide to Hobby Farming

Project Team
Editor: Amy Deputato
Copy Editor: Joann Woy
Design: Mary Ann Kahn
Index: Elizabeth Walker

i-5 PUBLISHING, LLC™
Chief Executive Officer: Mark Harris
Chief Financial Officer: Nicole Fabian
Chief Content Officer: June Kikuchi
Chief Digital Officer: Jennifer Black-Glover
Chief Marketing Officer: Beth Freeman Reynolds
General Manager, i-5 Press: Christopher Reggio
Art Director, i-5 Press: Mary Ann Kahn
Senior Editor, i-5 Press: Amy Deputato
Production Director: Laurie Panaggio
Production Manager: Jessica Jaensch

Library of Congress Cataloging-in-Publication Data
Ekarius, Carol, author.
 The essential guide to hobby farming : the how-to manual for creating a hobby farm / Carol Ekarius with Leslie J. Wyatt. -- 2nd edition.
       pages cm
 Title of first edition: Hobby farm: living your rural dream for pleasure and profit.
 Includes bibliographical references and index.
 ISBN 978-1-62008-144-0 (alk. paper)
 1. Agriculture--United States. 2. Farms, Small--United States. 3. Farm management--United States. I. Wyatt, Leslie J., author. II. Title.
 S501.2.E37 2015
 630.973--dc23
                        2015007137

This book has been published with the intent to provide accurate and authoritative information in regard to the subject matter within. While every precaution has been taken in the preparation of this book, the author and publisher expressly disclaim any responsibility for any errors, omissions, or adverse effects arising from the use or application of the information contained herein. The techniques and suggestions are used at the reader's discretion and are not to be considered a substitute for veterinary care. If you suspect a medical problem, consult your veterinarian.

i-5 Publishing, LLC™
3 Burroughs, Irvine, CA 92618
www.facebook.com/i5press
www.i5publishing.com

Printed and bound in China
15 16 17 18   1 3 5 7 9 8 6 4 2

Contents

Acknowledgments

> Go confidently in the direction of your
> dreams. Live the life you have imagined.
>
> — Henry David Thoreau

No book is the work of one person. Thanks first and foremost to Ken Woodard. He is my husband, my friend, and the one who makes everything else possible.

A number of people generously allowed me to interview them for the book and shared their personal triumphs and challenges. For their willingness to contribute their stories, I want to thank Dr. John Ikerd, David Muehleisen, Jill and Ken Giese, Jan and Tim Vala, Carol Ann Sayle and Larry Butler, Michele and Gustavo Huerta, Stephenie Caughlin, Judy and Sam Cavagnetto, Carol and Melvin Moon, Angel Henrie and Joseph Griffith, Susan and Stephen Robins, and Gary Dunn of the *Caretaker Gazette*.

Back to the Farm

When the United States formed "a more perfect Union, [to] establish Justice, insure domestic Tranquility, provide for the common defence, promote the general Welfare, and secure the Blessings of Liberty to ourselves and our Posterity,"* more than 90 percent of our population were farmers. They were the people who produced their own food and fiber, bartered for food, or bought food directly from someone else who produced it. Today, such a small percentage of our population is considered farmers that this group is truly a minority in our country.

At about the time Thomas Jefferson was penning the words of our Constitution, he wrote the following to President George Washington: "Agriculture…is our wisest pursuit, because it will in the end contribute most to real wealth, good morals, and happiness." Furthermore, "The moderate and sure income of husbandry begets permanent improvement, quiet life, and orderly conduct, both public and private."

Jefferson's agrarian ideal was not new; it was a viewpoint passed down by the earliest philosophers. As our population has increased and urban centers have grown apace over the past half-century or so, we've seen the agrarian ideal give way to an economic and social paradigm shift, resulting in fewer yet bigger *agribusinesses* supplying our food and fiber. Corporately controlled operations have steadily displaced midsized independent family farms, and those ex-farmers often end up moving to cities, seeking other means of livelihood. Iowa—the epitome of a farm state in many people's minds— provides a good example: It went through a landmark change some time

* From the Preamble to the Constitution of the United States

The USDA defines the different types of farms based on their annual income.

in the late 1950s, with more residents living in cities than on farms and in rural communities.

As remaining farmers grow their operations to try to stay in business, agriculturally induced environmental problems have increased. Runoff from agricultural production has contributed to polluted lakes and streams and is at least partially implicated in the ongoing issue of *hypoxia*, the lowering of oxygen levels in estuaries, waterways, and coastal waters such as the so-called dead zone found in the Gulf of Mexico.

Intermediate farms, smaller farming operations that are defined by the United States Department of Agriculture (USDA) as farms with less than $250,000 per year in gross farm income, have tried to remain in the commodities game but are not finding it easy. Statistics from the US Census Bureau show that the farmers in this earning range produce an average net income that hovers not much higher than 10 percent of their gross sales. It doesn't take a wizard with a calculator to figure out that, at this rate, a quarter-million dollars yields less than an optimal living wage. As a result, an overwhelming percentage of small farms depend on nonfarm income, including off-farm jobs, government subsidies, payments from programs such as the Conservation Reserve Program (CRP) and Grassland Reserve Program (GRP), and whatever other means of additional income they are able to obtain to keep their farms operational.

Agripreneurs

Some small farmers run their operations as successful commercial enterprises, not by competing with the big farms but by taking advantage of direct marketing, organic production, and other strategies to be self-supporting on a small acreage. Many do quite well, generating net incomes as high as 50 percent of their gross. These "agripreneurs" (agricultural entrepreneurs) build successful businesses on farms and ranches ranging from postage-stamp size to hundreds of acres. They're serious about making a profit, taking the initiative to learn about alternative production methods and making sure they are growing things they will be able sell, a requisite for making a profit. As experts will tell you, knowing your market is essential before putting anything on or in the ground!

For example, Melvin and Carol Moon, conventional berry growers in the Puyallup Valley in Washington, were selling to wholesalers, but they realized that they were going to need to do something different to survive. Thus, they launched Puyallup Valley Jam Factory, using their own berries to make jams. Their product took off, and they not only have customers lined up, waiting for their homegrown goodness-in-a-jar, they also ship worldwide.

Another success story is that of a professional couple, Stephen and Susan Robins, who retired to San Juan Island, north of Seattle. As Susan says, "We had twenty-five acres on the island that we loved, and we had to think of something to do with it. We wanted to preserve it for open space, so our first concept was that we would start an organic farm where we would have a crop that didn't use water, didn't use fertilizer, and where we could make added-value products on a year-round basis so we could spread the enterprise over the entire year instead of being seasonal."

Some farmers concentrate on a specific niche, such as growing organic produce, to earn a profit.

Their vision to protect a quiet valley from residential development and share that space with other islanders and visitors while making it productive and self-sustaining has developed into an organically grown lavender farm that includes an on-site distillery for essential oils, provides jobs for local residents, and produces handcrafted lavender products. Their products include lavender sugar, pepper, and vinegar; lavender soaps, shampoos, and body lotions; and lavender lip balm and massage oil.

Together with their crew, the Robinses grow and harvest the lavender, all by hand. Next, they distill the oil and create the products. Finally, they market the products at an on-farm store, at their local farmers' market, and through their website, www.pelindabalavender.com. Although they work with a few retailers who sell their products, Susan says, "We have discovered that our products sell best when they're in a closed environment—that is, a dedicated store—rather than sold among a lot of machine-made products. Our products are handcrafted and they are beautiful, but they get lost among the other ones that are slick." The direct marketing also enables them to capture a bigger portion of the consumer's dollar.

A Lifestyle Choice

Not everyone in the small-farm realm is ready for or interested in the kind of business that the Moons and the Robinses have developed. For example, Ken and Jill Giese raise a large garden on sixteen acres in New York, mainly to feed their family of six. They also raise turkeys and chickens, which they market directly or in cooperation with other small poultry producers. They didn't have much capital to get started, so while they were growing their business to a sustaining size, Ken worked part time on dairy farms and for construction companies in the area. The Gieses do much of their work with a draft horse, but when they need to use a tractor, they barter work for equipment use with some of the farmers Ken works for occasionally.

Judy and Sam Cavagnetto are also lifestyle farmers. Sam's job as a full-time over-the-road trucker for a large moving-van line allowed the couple to live anywhere they chose. They chose thirty-five acres in the mountains of Colorado. Judy and Sam wanted a great place with a small-town atmosphere in which to raise their kids. They keep horses, and the kids raise animals for 4-H.

Gustavo and Michelle Huerta also chose the country for its lifestyle. They were raised in Miami, but violence became too much a part of life there, so they decided to relocate to Tennessee. Here, Gustavo, a medical doctor, could establish his surgical practice, and the family could operate a small farm. With 200 acres, they raise a garden and horses, cattle, goats, and chickens.

Although these lifestyle farms don't have to make money for their owners, they pay with quality-of-life values that cannot be purchased. John Ikerd, Professor Emeritus of agricultural economics at the University of Missouri and a strong supporter of sustainable agriculture says, "What would you have to earn to buy the quality of life that a farm offers, from scenic areas and recreational opportunities to personal safety and a school for your children where the teachers know them and care about them? What is it worth if you are really living a life that has meaning in terms of a place to live? What would it cost for the view, the private schools, the clubs? In real economic terms, these are costs that shouldn't be marginalized."

For both the Cavagnetto and the Huerta families, it is lifestyle value that they seek. They believe that their children have experiences on the farms that they wouldn't get in the city or the suburbs. Indeed, making money is only one reason to farm, and, for many of these owners, not always the most important goal. Reflecting this evolution of focus, average farm size has dropped, reflecting an increase in small-farm operations in the 10- to 179-acre size range. Whether these farms have been in the family for a century or more or have been acquired by newcomers to rural life, small-farm owners demonstrate a variety of approaches to maintaining a diverse and vibrant agricultural community while protecting rural values, responding to consumer wants, and ensuring a healthy environment for generations to come. People are connecting with the land, with the seasons, with life around them. Thomas Jefferson would be pleased.

Small farmers reap the rewards of working the land and harvesting their homegrown bounty.

What Is a Hobby Farm?

The phrase *hobby farm* has become increasingly familiar in the new millennium. Although it is largely a tax-related term (more on that later), a hobby farm is basically a small farm that is operated for supplemental income or for pleasure. So how does this differ from the previously mentioned small farmers whose net sales are such that they often have to supplement their farm incomes through other means? In reality, not much. Most of the time, small farms are run by people who are trying to make a living from their property not by choice but by necessity, whereas hobby farmers tend to be people who choose to farm or live on a farm for reasons that may or may not include income. Whereas the term *hobby farmer* usually applies to those who leave more urban settings for their dreams of rural life, it applies equally

Hobby farming is often a family affair.

to someone who, for example, moved to a small Montana farm or ranch forty years ago to raise his family under the wide blue sky. Cattle sales paid the mortgage, and a garden provided healthy food, but putting gas in the tractor or clothes on the kids required him to work at an outside job. Hobby farming is not a new concept. It just never had an IRS designation before.

The Draws

Whether you made the move before it became the popular thing to do or have always dreamed of owning your own little corner of the earth, the draw of farm life is easy to understand. In our fast-paced, industrialized society, the farm harkens back to something simpler, quieter, and more in touch with our beginnings. It provides the chance to experience nature, in both her glory and her fury. In a society where urban and even suburban living reduces necessary chores to housekeeping, making sure the trash cans make it to the curb, and possibly some yard work or cleaning the pool, farms by their very nature offer many more real-time opportunities for families to work and "do life" together.

As Jill Giese says, "The benefit to us is, as a family, we can work together and play together. Our kids have learned the rhythms of helping; they know that animals need to be fed and watered, just like they [themselves] need to eat and drink. They're seeing how you spend the money to feed the animals, and you get the money back when you sell them. They each have some work in the afternoon: somebody feeds the turkeys, somebody does the watering. So they see responsibility and that we are depending on them and trusting them. They make their own games and run all over... using a pile of hay bales to play king of the hill. And they know about life, about how the male goat breeds the female goat, and that that's a natural part of life. They have seen baby goats being born, and this year we did a hatchery and had chicks coming out of their shells; they learned about the formation of the chicks and how different reproduction is from mammals to birds. It was exciting after three weeks, when the chicks came out, so Ken had them draw pictures and write things about the experience."

Many Baby Boomers are retiring early—often in their early to mid-fifties—and the farm offers them the chance to stay vibrant and learn new skills while they capitalize on their expertise and talents. Susan Robins says of their venture, "We weren't insecure about the business side of it, but we just were not experienced in the horticulture and product sides of it. It took over a year of intense research to get ready." However, now, with the business up and running, the couple takes great satisfaction in the outcome: "We've been gratified to have created an industry on the island, in a place where that is very hard to do, and we have created wonderful jobs for artisans and others who want to be involved in it. We think we have created a landmark on the island that is open, free of charge, to anybody who wants to come, and we get a huge amount of pleasure out of those people that come and picnic on the farm, bringing their kids and sitting in the fields when they're in bloom."

Farming also offers the chance to "experience the unusual or unexpected little adventure that lightens and even makes gladsome the work," as Gene Logsdon, a small farmer from Upper Sandusky, Ohio,

Children can share in many of the responsibilities of farm life.

points out. He shares some of his own adventures, such as "a pale green luna moth fluttering in the porch light, a fungus that looks like a little pile of sand; an ant milking its own herd of aphids; a killdeer nest right in the middle of our gravel driveway. And three years after we planted paw-paw trees, the first gorgeous zebra swallowtail butterfly, which feeds only on paw-paw, landed daintily on the tractor."

These "little adventures" may seem inconsequential, but for those of us drawn to the farm, they are grand payment for our labors. I have always been grounded by nature, and I love my own little adventures. As I broke the ice off the stock tank this morning, I smiled at a donkey nuzzling my coat pocket to find a treat, watched the mountain chickadees skipping in and out of the dried leaves of the currant bushes, smelled the fine scent of pine smoke in the air from our woodstove, and listened to the wind singing through the trees. In the summer, I can sit in an aspen grove and watch adult great horned owls teach their young to hunt or enjoy the raucous fighting of a small flock of pygmy nuthatches as they bathe in the puddle where we empty the stock tank.

The Challenges

Yes, life on the farm has magical moments, but as gratifying as farm life can be, reality is that it involves a level of physical work that many escapees from urban and suburban America have never engaged in, and it often necessitates adopting new working patterns. For example, if you have livestock to take care of—feed, water, milk, gather eggs from—vacations, holidays, and weekend trips become much more challenging to plan. If you're growing crops, you'll need to get used to dawn-to-dark days during planting and harvesting. Dreaming of running a farmstead B&B? Make sure that you are prepared for the lack of privacy that such a venture inevitably brings.

Our dreams of farm life somehow never seem to include such things as mud and mess and midnight vigils with some farm-related crisis or another. And make no mistake, rural living also means that you're going to have to drive a certain distance to do *anything* off the farm. Grocery shop. Soccer practice. Haircuts or nail appointments. School events for the kids. Whether you're taking in a movie, hankering for a specialty coffee shop, or meeting friends for dinner or a play, it will entail driving. That's the long and short of it. The miles and the minutes add up. You might save money by growing your own produce, but you'll spend more in gas and time. It's not a deal-breaker, but it's something to be aware of.

By its very nature, moving into a new neighborhood means becoming acquainted with the others who live there, and this is true in rural settings, too. Integrating into a community—rural, urban, or otherwise—means investing yourself in coexisting with, interacting with, respecting, and helping your neighbors, whether they live in the apartment above you or half a mile down the gravel road.

In areas where farms are passed down through families, it can be easy to feel like a newcomer. But don't make the mistake of stereotyping long-term residents. Few farmers will shut you out if you treat them with respect rather than condescension. No one appreciates someone who acts as if long-time country dwellers don't care about the finer things in life or have progressive ideas. Humbleness and a genuine regard for who they are, what they know about farming and country living, and the sheer amount of hard work they do goes a long, long way toward making a place for yourself in the neighborhood.

Depending on the area to which you move, it may be hard to shed the newcomer label. When we first moved to our farm in Minnesota, we were having coffee one day with our neighbors, Bev and Willy. We had been able to break the ice quicker with them than with any of the other neighbors because Willy had been renting the fields when we bought the place, so we had an immediate business relationship with him. In their sixties at the time, the couple lived on the farm that Willy's grandfather had homesteaded almost a hundred years earlier. We asked if the neighbors on the other side of them, Kathy and Jim, were originally from around the area. "Oh no," Willie said, with a shake of his head, in a way that clearly indicated that Kathy and Jim were like us—foreigners. "They are from over near Deer Creek." Deer Creek was only about twenty miles away, and despite the fact that they had lived on their farm for well over a decade, Kathy and Jim were still not locals.

Don't take this personally. It's a dynamic that takes place everywhere in the world, not just in rural areas. People have to get to know and trust each other, and the longer you live around someone, the better you get to know him or her. It takes some real interactions, such as helping each other in hard times or sharing the bounty from your garden or oven—along with friendly conversation—to start to know your farm neighbors and for them to get to know you as well.

A spacious suburban backyard can serve as a viable hobby farm.

Do you want to start your hobby farm in a rural area or somewhere more suburban? Consider the pros and cons of each.

When a tornado took the roof off a neighbor's barn in rural Missouri, one family of fairly new hobby farmers joined in helping pick up sheets of tin roofing out of the adjoining fields and brought food to the family that had been hit, and, in the midst of loss, their friendship took a leap forward.

Another instance with the same hobby-farm family arose from a calving crisis with their newly acquired Jersey milk cow. A quick call to the neighbors down the road, and within minutes came the farmer, his wife, and their little granddaughter. Not only did they help deliver a healthy calf, but they also donated a bale of prime hay that they'd just cut, and, for the next few days, they drove up to check on the cow with their granddaughter in tow so she could visit the calf and some newly hatched chicks.

Stopping by for a chat in the driveway on the way to town or waving to each other as you pass—every interaction is a step toward friendship. Life together is built of such exchanges, and while you may never be viewed as a "local" of the area, you can develop some deep and long-lasting relationships built on mutual trust and respect.

There is a certain neighborly charm and support in farming communities that's often missing in urban and suburban America, as well as a generosity that often exceeds their means. They'll plow your driveway after a snowstorm. Help you fix your fence when your cows get out. Help you troubleshoot your air conditioner, your lawn mower, or your cistern pump. There's something solid and enduring about those who live on the land. They don't fight nature—they work with it and may bow under it, but they do not break. Invest some time, effort, and kindness into forming relationships with your neighbors, and you'll soon be rewarded with a sense of belonging that goes well beyond labels.

Hot Spots for Hobby Farming

Most new small farms are in the "rurban" (that interface between urban and rural) zone: away from town, but not too far "out in the boondocks." Rurban areas boast good roads and proximity to large groups of consumers for direct marketing. They benefited from the burst of technological innovations in communications and computing during the 1980s and early 1990s that enabled more people to work from home. These features make rurban communities most attractive for those seeking small-farm life. And because of the increase in small farms, there has been an increase in services—such as cooperative-extension programs and specialty equipment and tractor dealers—designed to meet the needs of hobby farmers. Although the majority of viable small farms are in rurban areas, other rural areas are beginning to see the transition as well. This is in part the result of ex-urbanites taking advantage of lower priced farmland that's found in truly rural communities, but it's also being helped by increasing numbers of farmers' markets outside of urban centers, together with programs sponsored by the USDA that allow people with lower income to use food stamps and federal coupons at farmers' markets. According to USDA research, the number of farmers' markets in the United States has grown dramatically. These markets provide a crucial source of revenue for small farmers from California to Missouri and beyond.

You don't need acres of land to set up productive growing spaces.

Last, a growing number of farms are located right in urban core areas and near suburban residential areas. Today, substantial agricultural production in the United States originates from within metropolitan areas. Some city farms are community collaboratives, sprouting up in some of the poorest and most violent neighborhoods in America through the efforts of nonprofit organizations trying to improve the quality of food and provide positive opportunities for inner-city dwellers. Small-scale agripreneurs run other city farms, seeking viable businesses in the places in which they live. Typically operated on less than five acres and using the community-supported agriculture (CSA) model, these farmers are taking advantage of their ability to build close relationships with the families who purchase from them.

Since 1991, Larry Butler and Carol Anne Sayle have built Boggy Creek Farm in East Austin, Texas, into a very successful urban farm. With five acres and a 160-year-old farmhouse located just blocks from the Texas state capitol, they market a wide variety of grown-on-site produce and flowers; value-added products, such as smoke-dried tomatoes and salsa, created from their surplus; and animal products, such as goat cheese and free-range eggs, which other farmers in the region raise.

"I was getting burned out remodeling houses and selling real estate," Larry recalls. "We started growing vegetables on some land we owned that's about an hour and a half outside of town. Once we grew all this stuff, it was like, 'Well, now what are we going to do with it?' A friend of mine had a liquor store here in Austin and let me sell from a card table in front of the store on Saturday mornings. The first Saturday, I made about forty dollars on carrots and onions and greens. The rest of the story is [that] it got nuttier every week, and now we are running four cash registers."

In 1992, they saw the Austin property, complete with nut trees, irrigation water, and a historic house, listed in a real-estate book. Ninety days later, they were signing the papers and launching into full-time farming. They haven't looked back and have no regrets.

Rurban

Rurban communities are those that offer a rural lifestyle and still have traditional rural populations but are strongly influenced by urban areas. They are usually located within a couple of hours' drive from a major metropolitan area or a major resort community. They may have a fair population of telecommuters who occasionally go into the city for work, but rurban areas are just a little too far out for many day-to-day commuters.

Are You Ready for the Country?

I love this life. I wouldn't trade it for a fashion model's body, a vault full of money, or a best-selling book. How could you want to trade something that fits so well, like a treasured old coat that warms you to the soul? I love our compact house, nestled in a saddle between two pine-dotted ridges, overlooking Pikes Peak to the east and the Continental Divide to the west. I love that our house is charged (electrically) by rays of the sun bouncing off sparkly blue solar panels and is heated by wood from nearby beetle-killed trees, which my husband, Ken, brings home with the donkeys or with an old three-wheeler and a trailer. I love the fact that I can look out my window and see animals—both our own animals and wild ones—gamboling about, healthy and happy, as animals should be. I love the new potatoes that come from the garden and the greens that grow in pots in the living room that freshen our winter plates. I love to sit in the evening and listen to Ken serenade me with his guitar, both with wonderful songs he has written and covers of some of my favorites. I can go to the city if I need a city-type fix, but I hardly ever do. Ken and I mostly stay home, both of us working from right here, although occasionally we go to one of several nearby small towns for shopping or a night out.

But as much as I love my life, I have to be truthful: it can be really challenging, and I know that it isn't for everyone. For example, as I write this chapter, we are coping with a frozen water service line (our first at this house after six winters, though, sadly, not our first ever—and not likely to be our last). Yesterday, we spent our Saturday fighting the water line, thawing snow on the woodstove to keep the critters watered, and unloading

Living in a rural area means being prepared for all seasons.

our midwinter delivery of hay, all while Ken was combating a cold. Since we were unable to get the line thawed, we went to bed dirty. Luckily, it was a sunny and pleasant day, but it was the tail end of a three-week spell with night temperatures dropping close to –30 degrees Fahrenheit.

We have had other bad times, as have most people who relocate to a farm from a suburban or urban life, and those bad times should give you pause. Moving to a farm is a family lifestyle change, and the whole family needs to want to face a unique set of challenges that don't exist for most urban Americans today. If your wife's definition of a challenge is picking out a shade of nail polish that will match her silk blouse, and her idea of fun is spending a whole day cruising through Saks Fifth Avenue, she probably won't be happy on the farm long-term. If your husband's idea of hard physical work is mowing the 100-square-foot lawn on a Saturday morning, and he is quite proud of the fact that he has never had a blister, he may not be ready for how many hours of physical labor he'll have to put in on a farm. If your kids don't spend a lot of time outside, they might not feel the need for the fresh air and sunshine of farm life, and it's pretty safe to say that the novelty of chores—feeding, watering, cleaning, milking—wears off at a speed in direct proportion to the age of the child. Little ones look on it as "playing adult," but teens tend to view the manual labor, smells, and day-in/day-out regularity of chores as something akin to modern-day torture. However, also know that they will gain and utilize skills for which they may perhaps one day even thank you.

If in doubt about the family's sincere commitment to the lifestyle change, consider practicing farming in your own backyard before you make a big move. The potential to grow a food-supplementing garden exists in most places. Many cities even have community gardens where residents who lack an adequate yard for a garden can acquire a small bed to work. A few backyard chickens (minus a rooster to wake up the neighborhood) are a great addition to most yards as long as local zoning laws permit chicken keeping.

There are also some other opportunities to wet your toes without drowning. *Agritourism*, or vacationing on a working farm or ranch, provides a temporary taste of the lifestyle. Find agritourism opportunities in areas where you think you might like to relocate by contacting that state's department of tourism or by going online to www.farmstop.com.

Many working farms also offer opportunities to work as an intern, and if there is a community-supported agriculture (CSA) facility near you, you can usually opt to participate in farm activities such as planting, tending, and picking produce and/or animal husbandry if you become a member of the CSA. Check out websites such as the National Sustainable Agriculture Information Service at attra.ncat.org or do an Internet search for "organic volunteers," which will turn up such things as World Wide Opportunities on Organic Farms (WWOOF) and a multitude of similar opportunities.

Another approach to testing the farming life without making a final commitment is to work as a caretaker for a farmer, rancher, or rural landowner. According to the editor of the *Caretaker*

Gazette, there are thousands of opportunities for property caretakers. Some caretaking prospects are relatively short-term commitments, like farm-sitting for a month, but others are long-term arrangements that can go on for years. Since 1983, the *Caretaker Gazette* has listed opportunities for property caretaking around the world, and it gives readers the option of placing situation-wanted ads as well. Visit the *Caretaker Gazette* for more information at www.caretaker.org.

When in Rome

One of the first things to think about with your move to the country is that you are opting for a change in culture. The further you move from the urban fringe, the wider the cultural differences may seem. These dissimilarities are not insurmountable, though. Remember the adage: When in Rome, do as the Romans do.

Focusing on the differences between the lifestyle you leave and the one you step into can make for a difficult transition. The problems you'll encounter in that realm depend for the most part on how you view and interact with your new neighbors. Our experience in Kremmling, Colorado, provides a perfect example. We found ourselves neighbors to a codgerish old fellow (probably not much older than we are now) who ran a junkyard and who seemed distinctly unfriendly at first. Turns out that the people who had built the house we were living in tried to get his junkyard (which was there long before they bought the land and built the house) closed down by the county. Erwin suspected that we were more of that ilk—city slickers trying to change the countryside.

The rancher who owned a large expanse of land up behind ours began dropping cows and calves (hundreds of them at a time) down at the bottom of the valley, and he, along with his kids and hired hands, drove them up through our place to get to his land.

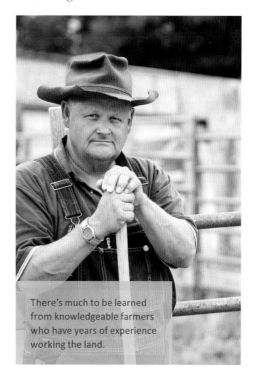

There's much to be learned from knowledgeable farmers who have years of experience working the land.

Like Erwin, George, too, had had a run-in with the previous residents, who had hassled him about moving the cows through and left gates open when they were supposed to be closed or closed when they were supposed to be open. He eyed us with suspicion that first spring morning.

It turned out to be fairly easy to break the ice with these fellow country dwellers. We smiled at them when we saw them and followed the rural tradition of waving when we passed them on the road. We tried hard not to judge them by our "citified" criteria, and, before long, we were on good terms. They began visiting us and would offer advice and help. By late summer, George had given us permission to ride our horses on his ranch, which ran for tens of thousands of acres up behind our house.

More often than not, hard work and hard economies have shaped your new neighbors. They are dedicated to their land and their family, which is often an extended network within the community. They value tradition in work and often in worship as well. Older neighbors in particular may be uncomfortable with such things as hammering out the details of a business deal in a contract (they tend to stick to the handshake approach of doing things) or in talking about business with a woman. Accept and value your neighbors for who they are, and they will be more likely to do the same for you.

Economics of Farm Life

People coming from towns or cities soon discover that although real estate in the country is somewhat cheaper than comparable real estate in developed areas, farm living isn't automatically cheaper than life in the city or suburbs. Rock stars and corporate executives looking to get away from the craziness of their day-to-day lives don't have to think much about the money, but if you don't fall into the multimillionaire category, you may need to consciously make changes in your spending habits.

The job market in rural and rurban areas is often limited, and the pay scale can be lower than those found in cities. You may be able to commute into a city for work, especially if you have the ability to arrange flexible scheduling with your employer, but you should still think about what commuting long distances will mean to your life and your family. Technology has made telecommuting possible for many people, but some rural areas still have limited access to the Internet—and don't expect high-speed service. Many rural communities have not yet been blessed with digital subscriber line (DSL), so if your at-home business requires a fast Internet connection, you'll want to research network speeds.

Cell phone coverage is another issue because it's more difficult for some carriers to transmit signals in certain areas. Even if poor cellular coverage is not a deal-breaker in the purchase of your dream farm, it's nice to know what you're up against before you back the moving truck up to the door.

Certain costs of living may be lower in rural and rurban areas, but remember to consider the costs of any equipment you'll need to maintain your farm.

Sustainability and Self-Sufficiency

Farm living can be just as comfortable as city living, but sustainability and self-sufficiency are the pillars of farm life. You can raise at least some of your own food and reduce your dependence on outside resources that eat up your money. Your entertainment can be right in your own backyard.

Michelle Huerta from Tennessee says that when her family moved to the farm, they had both pleasant surprises and sticker shock about the economic differences between the city and the farm. "The first thing that struck us was that we could go out to eat a lot more often because the restaurants are so much cheaper here. The six of us could go out for a meal for [around] thirty dollars. At the same time, it is very expensive having a farm; there is no doubt about it. It's shocking to look at the price of a tractor or farm equipment, and a lot of services, like having a plumber or an appliance repairman out, are much more expensive here than in the city."

Selling your produce at farmers' markets or right from your farm can be profitable.

Maybe your plan is to create a moneymaking enterprise out of your farm—to be a real farmer. You can make money off of a small farm, but it is hard work. If you aren't starting out with some cushion of money in the bank, things can careen into a financial wreck before you know what happened. For planning purposes, assume that you won't make money on your enterprise for at least the first couple of years. Can you get by with outlay and more outlay when there is no significant farm-generated income?

Farms and the IRS

Our accountant once told us that the surest way to find ourselves at the uncomfortable end of an IRS audit is to show a farming loss while showing significant off-farm income. The IRS considers that if farming is a business, it must be set up and run with the intention of earning a real profit. When a farm is set up and run as a profit-making endeavor, you can deduct the expenses of running the business, even if your expenses exceed your income and you have a loss—and the loss may be applied to other nonfarm income under the federal tax code. But the IRS is also wise to the fact that many people set up farms with no serious intention of making profits, yet they deduct those "farming losses" from their income. If the IRS determines that your farm is an "activity not engaged in for profit," then you fall under section 183 of the tax code, which governs "hobby losses" (this is the IRS term that gave rise to the label "hobby farm").

Losses incurred in connection with a hobby are generally deductible only to the extent of the income produced by the hobby. In other words, you can't use a hobby to generate a tax loss that shelters your other income. And, all of your hobby income is supposed to be reported as "other income" on line 21 of your form 1040 income tax return, although the expenses incurred earning that income are deductible only if you itemize your deductions (they are considered "miscellaneous itemized deductions"). You can only deduct the portion of the expenses associated with your hobby income that exceeds 2 percent of your adjusted gross income.

When determining whether you are operating a business, as opposed to engaging in a hobby, the IRS considers the following points:

- The manner in which the taxpayer conducts the activity. Do you operate in a businesslike manner? Do you keep complete books and a separate business checking account? Do you advertise your products? Do you study what is going on in your industry and adopt new techniques to increase profits?

- **The expertise of the taxpayer or his or her advisers.** What is your background in the activity (including number of years of formal training and practical experience)? Do you consult with professionals, such as attorneys or CPAs? Do you seek advice from experts, such as county extension personnel? Can you demonstrate that you have acted on the advice offered by your advisers? For example, have you fertilized a pasture based on the extension agent's recommendation?
- **The time and effort that the taxpayer spends on the activity.** Do you spend a significant amount of time on the business and its related activities? Do you actively spend time marketing your products? Do you keep any kind of time log that documents your work and efforts to make this a profitable operation?
- **An expectation that assets used in the activity may appreciate in value.** Do you anticipate that certain activities, like breeding purebred stock, will show a profit down the line through appreciation? Are you in an area where actively farmed land is going up in value faster than other land? Can you demonstrate that your efforts—for example, a program you have undertaken to reclaim your land from a noxious weed—will increase the value of your land?
- **The taxpayer's success in similar activities.** Have you started other similar enterprises in which you did show a profit? For example, if you started and ran a successful kennel, you demonstrated the ability to create success for profit. The IRS doesn't consider success in your primary field as proof. In other words, it doesn't count for you to have been a successful banker, but if you were a banker who started a profitable restaurant as a secondary business, the IRS is more likely to concede that you are in farming for a positive business result and not just a hobby.
- **The taxpayer's history of income or losses with respect to the activity.** Did you have a profitable farm at some other point in your life? Has your bottom line been steadily improving over several years and your losses getting smaller?
- **The amount of profit from the activity.** Have you had some profitable years already on the farm (even if the profit was slight)? The profit has to be legitimate farm-generated income and not from an unrelated activity.
- **The taxpayer's financial status.** Where does your income come from? How much money do you have from other sources? Is the farm your only source of income?
- **Elements of personal pleasure or recreation.** This is probably the most interesting factor because even the IRS acknowledges that you can derive personal pleasure, beyond the profit itself, from an activity that is intended to make a profit. Yet the IRS expects that the pleasure is secondary to the profit motive.

Searching for the Ideal Farm

Farm shopping is fun, but it is best to take your time. First, make a list of what you want in a farm and your general preferences for variables such as climate, terrain (mountains, seashores, plains, forests), what you want to grow and/or raise on your farm, cultural resources, churches, schools, and hospitals, as well as proximity to your extended family, the job market, and markets for farm products you are thinking about raising. This list will initially help you target areas that meet your needs and will later help you narrow the field of farms in the area in which you're interested. Once

A Wisconsin farm under a beautiful blue sky.

you identify areas or regions of the country that interest you, try to find out as much about them as you can. The Internet makes much of this kind of investigation easy. You can also research by calling state economic development and tourism offices, town and county clerks, county extension agents, school districts, and/or chambers of commerce. Realize that local personnel may not be unbiased sources, but they're the ones who really know the good, the bad, and the ugly of a community.

Assess how the area is doing economically—growing, stagnant, or in decline. If you have school-aged children who will be attending the public schools, research the school districts and ask questions about graduation rates and standardized test scores.

How are the local healthcare facilities? Find out about the quality of healthcare organizations by contacting the Joint Commission (JCAHO), which accredits healthcare organizations. The Joint Commission provides quality reports on most hospitals and surgical centers, nursing homes, home care services, and other types of healthcare operations. Their website, www.jcaho.org, offers a searchable database of reports, and you can purchase reports from them as well.

If you are not in a big hurry to move, consider subscribing to local newspapers of those communities that interest you. Contact numerous real estate agents in the areas that seem to meet your general preferences; they will be happy to send you information on the farms they have listed. By dealing with multiple real estate agents in the same vicinity, you'll get a better range of properties and prices. United Country Real Estate has the largest network of rural real estate agents in the country. They publish a catalog that you can request, or you can search their online database at www.unitedcountry.com.

As you narrow your choices, plan some trips to visit the top two or three areas you have identified. While there, look at farms that are for sale, but try to resist the temptation to make an offer until you have really gotten to know each area. If possible, visit again during a different season, even if only for a short stay.

Serious farm shopping comes after you have settled on an area. Again, your initial criteria and questions will help you compare different farms, but, for this phase, you should do some preplanning to help you do an even more thorough analysis of the individual properties that interest you. We use a technique known as *matrix analysis* when we shop for real estate (or any other major investment). Put simply, with this approach, you set up a worksheet that allows you to "compare apples to apples." Write down the criteria on which you will judge the properties, and establish a scoring system for each criterion. Points to consider on your matrix may include the broad categories, such as quality of schools, healthcare, and recreation. But, more importantly, remember to look at the specifics. What type of soil does the farm have? Is the land in pasture or field crops? Does the size and condition of the house and outbuildings meet your needs? If not, what will it cost to make repairs or upgrades? Are there land amenities, such as a live stream or a nice old hardwood forest? Does the well produce sufficient water and is the quality good? Does the septic tank work? Are the utilities suitable to meet your needs? How far are the buildings from roads and neighbors? Four of the most important issues to consider are water, drainage, septic systems, and utilities.

Water

I think water is the most important single issue when shopping for a farm. People living in towns and cities generally have municipal water—turn on the tap, and out comes good, safe water—but

Your chosen property must have an adequate supply of good-quality water for your crops, your animals, and your family.

farms are almost always on private wells. Some may have cisterns to which water must be transported, either by hauling it from a community well or from some other water source. One Missouri farmhouse dating from 1885 operated on an archaic arrangement of well water pumped into the original cistern from which a shallow well pump pulled water into a pressure tank under the house. The realtor assured buyers of high-quality water, and it did taste great, but when they ran a water test after purchasing the farm, they found that the cistern leaked and that *E. coli* was an issue. Modernizing the system required a considerable financial outlay.

Your first concern, therefore, is water. Ask to see a current water test and well production report. A good well produces adequate quantities of safe water for your family and animals to drink and for necessary irrigation of gardens and crops.

Quantity may not be a problem if you are looking in a high-precipitation area of the country, but if you are searching in the arid West, it could be a major consideration. Judy Cavagnetto says that she and Sam learned this the hard way. "We had our house built and then moved from Wisconsin [to Colorado]. We ended up having to have a cistern put in the basement because the well wasn't producing enough water. That's something we found out the day we moved here, when I went to water the twelve horses [and] there wasn't enough water; we could only get about fifty gallons. The drillers drilled to 250 feet and got water, but it was the minimum gallons per minute for a well. Now I wish they would have drilled deeper so we would have been closer to the maximum than the minimum."

Check with neighbors, local well drillers, the county, or the state agency responsible for water resources to find out if there have been any dry wells or wells with poor production in the area. You can hire a well driller to test a well if there is any reason for possible concern. When purchasing bare land (especially in the West), add a condition to the contract that says you have a certain period of time before you buy to drill a well that gets sufficient water by a certain depth (well drillers, or the state water resources department, should be able to give you guidance on how deep that is). If the well driller can't hit water, you are out the drilling cost, but you aren't stuck with a piece of dry land.

Make it a condition of the contract for purchasing your farm that the well meets all standards set forth in the Environmental Protection Agency (EPA)'s Safe Drinking Water Act. If the well doesn't meet these standards, you don't want to buy the place. Have your water tested for herbicides, pesticides, and nutrients, as well as bacterial contamination. Contact the county or state health department for recommended labs.

Don't forget to taste the water because not all water is created equal, so to speak. Water carries varying minerals depending on what part of the country and which aquifer it comes from. High iron content will leave rust marks on plumbing fixtures unless a water-treatment system is installed. Mineral content also affects the water's flavor. For example, there is an aquifer identified as the Roubidoux (called the Jordan Creek locally) that runs under Missouri and states south and east. Although plentiful, water from this strata often carries sulfur. This means nothing to someone who has never tried to drink the stuff! Although it will not hurt you or your livestock and gardens, the smell and taste leave much to be desired (although those who drink it untreated report that mosquitoes never bite them). If the farm you're investigating has a sulfur well, and it doesn't already have some method of outgassing to render the water drinkable, you'll need to install one. Local water treatment companies can help you explore options.

Some areas of the country have what locals refer to as "soda" water, which contains a sort of bicarbonate that gives the water the taste of baking soda and causes it to hiss and spit as it first comes through the water lines. Such water is very "soft," meaning that it takes very little soap to make suds as compared to "hard" water, which contains high levels of calcium and magnesium. Again, there are home water treatment systems available to correct mineral balance. What you want to avoid are unpleasant and costly surprises resulting from overlooking the existing water situation.

A farm that has a river or creek running through it offers aesthetic and recreational value, but it may be subject to flooding, and just because it hasn't flooded yet, doesn't mean that it won't. Look around you. Buildings should be located away from the banks of the river or creek and on high ground. The Federal Emergency Management Agency (FEMA) is the national agency on flooding, and it often has flood maps available. If you are in a FEMA-designated flood zone, you'll want to check into what flood insurance will cost. For farms in floodplain areas, most lending institutions will require buyers to purchase flood insurance in order to obtain a loan.

Finally, is there the right amount of water for the kind of farming you plan to do? In some areas,

Look for wet or swampy areas on the property.

crops need irrigation. In other areas, swampy pastures grow poor-quality forage. County extension personnel are the best folks to provide information on water needs and availability for crops in the area you are looking at.

Drainage

Life in a swamp can be miserable for both you and your critters. Over the years, we've known a number of neophyte farmers who bought farms during a dry spell only to have their dreams drown in knee-deep mud when the rains returned. Their crops weren't good, and their animals suffered from myriad health problems, such as hoof rot and mastitis (an infection of the mammary gland found primarily in lactating animals). Their equipment spent more time stuck in the fields than operating, and their buildings had water-related problems, including wet basements. Although you can usually correct minor drainage problems through the use of French drains and grading (labor and costs should be considered), it is best to avoid low-lying areas.

Even if you are shopping for a farm during a drought, you can assess potential drainage problems in several ways. The first, and probably easiest, option is to talk to people in the area about the land. Locals will know.

Another option is to study a topographic map of the area. The United States Geological Survey (USGS) prepares these maps, which display a wealth of information about elevations, waterways, and vegetation. A topographic map will display marshy areas by showing blue dashes and blue vegetation. Also look for intermittent lake beds, which are displayed by a dashed blue line with light blue lines within its boundary. You can purchase topographic maps at stores that cater to hunters, fishermen, hikers, and other outdoor enthusiasts or through the USGS Map Store (http://store.usgs. gov or 800-ASK-USGS). Additionally, with Google Earth, you can see an aerial view of the land, which reveals vegetation and terrain very well, even if water is low at the time of the photo.

The third approach is to study vegetation. Plants that grow in wet ground have adapted to having their "feet" wet and tend to have thick, waxy-surfaced leaves.

Septic Systems

A civil engineer or health department employee designs standard, modern septic systems according to uniform design specifications. Each system consists of a tank and a drain field (also called a leach field or absorption field). These systems efficiently treat wastewater, thereby protecting groundwater, rivers and lakes, and drinking-water wells. However, in some areas of the country or for certain types of uses (such as business uses), systems may have additional components, such as effluent filters, surge tanks, grease traps, and/or lagoons. These additional requirements are sometimes

Maximizing Your Septic System

Here are four steps to help maximize the life and efficiency of your septic system and to protect the environment:

Inspect regularly. Have the system inspected on a regular basis (every two to three years). This extends the life of a septic system and helps avoid unnecessary and expensive repair and replacement costs.

Conserve water. Hydraulic overload is a major cause of septic-system failure. Low-flow plumbing fixtures, faucets, and showerheads will minimize the amount of water that enters the septic system. Stagger water use throughout the day to minimize the strain on your septic system.

Protect the drain field. Do not plant trees and shrubs with deep, penetrating roots near the drain field because the roots can plug the perforated pipe structure. Do not drive or park vehicles and equipment over the drain field because their weight can compact the soil and damage drain-field components.

Manage household waste. Limit the types and amounts of waste poured down the drain. A garbage disposal can double the amount of solids added to the septic system, so use a disposal sparingly, if at all. Opt for composting, disposing of waste in the trash can, or giving edible scraps and leftovers to the chickens. Cooking oils and fats harden after disposal and block the septic tank inlet, or outlet, and they can even clog the soil pores surrounding the drain field, reducing its effectiveness for filtering wastewater. In addition, never dump chemicals like paints, solvents, drugs, and pesticides down the drain because these items may kill the microorganisms that help purify wastewater, and they can potentially enter into the groundwater and contaminate the drinking-water supply. Using low-phosphate detergents can reduce phosphorous loads to surrounding lakes and streams by as much as 40 percent.

needed where geology limits the effectiveness of drain fields or where water quality needs to be protected at a higher level— for example, near small streams with cold-water fish populations or where the density of development is expected to affect nearby lakes.

The designer of the system sizes the components according to estimated usage. A two-bedroom house will have a smaller system than a five-bedroom house, and a shop will have a smaller system than a restaurant.

Most modern tanks are concrete, although, in some areas, local authorities may permit the use of fiberglass or polymer tanks. Older tanks were often made of steel, which rusts through after twenty to forty years of use. (If you purchase a really old farm, the "tank" may actually be a buried car body.) In the tank, solids settle to the bottom and partially decompose through a biological process known as *digestion*, and scum from soaps and oils floats to the top.

The liquid in the middle exits the tank and flows through a series of perforated pipes and out the drain field. The soil is the final and most important component of a septic system. This is where the majority of wastewater treatment actually occurs. Through physical and biological processes, the soil consumes most of the bacteria and viruses, as well as some nutrients, in the wastewater as the wastewater effluent travels down through the soil layers.

To maintain efficient operations, you need to pump the tank from time to time. Pumping removes the solids from the bottom and the scum from the top. If not removed occasionally, the solids and scum begin to take up the entire volume of the tank. Once this occurs, they move into the drain field, quickly clogging the soil matrix that is critical to treatment. Usually, when this happens,

Solar panels keep utility costs down and are kind to the environment.

you need to dig up and rebuild the drain field, which is far more expensive than keeping up with pumping.

Utilities

Most towns and cities have readily available utility services for electricity, natural gas, phone, and even cable television; wires and pipes run into the house, and essentially all you have to do is pay the bill. Farmers may or may not have these services that townies take for granted. Few rural areas have cable television or DSL internet, and natural gas rarely runs to farms.

We live "off-the-grid," in a house powered by photovoltaic solar panels and with a gasoline generator for operating our water pump. Off-the-grid living is great, but developing an alternative-powered home requires a little more planning and initial expense; however, you will never see an electric bill again, and you can laugh when your city friends complain about power outages. You will be helping to protect the environment by reducing nitrous oxide, sulfur dioxide, and carbon dioxide emissions. You will need to be more diligent in turning off unused lights and appliances, but it seems a small sacrifice for the benefits you reap. We have six solar panels, four batteries, and an inverter, which run the lights, television, stereo, computer, and other small electronic devices. To reduce the demand for electrical power, all of our major appliances run on propane, including the refrigerator. (Electric refrigerators and freezers draw a lot of power, so you will need a significantly larger system to use them.)

Our water pump is fairly deep, so we opted for a gasoline generator to run it, which cut down the capital expense. Down the line, however, we may redesign our water system to eliminate the generator. We run the generator once or twice a day.

For those considering off-the-grid living, I'd recommend solar as the first choice in areas of the country that are blessed with plenty of sunshine. Wind generators will often do the job in areas that lack enough sunshine to make solar a viable power supply, but they have moving parts, which means wear and tear, maintenance, and repairs, and some models can be kind of noisy. Solar panels, on the other hand, are virtually maintenance-free and absolutely quiet.

Small hydro systems may be an option if you have live water in the form of a stream or river running through your property. Hydro systems use one of man's oldest inventions—the waterwheel—to turn a generator turbine. Hydro applications work best where there is a good drop along the river's course, which creates *head* that turns the wheel. Three feet is about the minimum head needed to produce electricity.

Wind is another alternative source of power sometimes used by farmers.

Whichever off-the-grid system you choose, you'll need to perform routine maintenance on your batteries. Ken checks our batteries twice a year, cleaning terminals, replacing corroded battery cables, and checking water levels in the batteries.

Real Estate Legalities

It is best to have someone highly knowledgeable to represent you in the state where the transaction is taking place. This could be a real estate agent or an attorney, but keep in mind that if you are looking at real estate with a real estate agent who has a listing on the property, he or she is legally representing the interests of the seller, not yours. No matter how knowledgeable your representative, though, it is important to educate yourself on real estate legalities. Be aware, for instance, that there are different types of deeds—with very different warranties and securities—and land interests and codes that could affect your future plans.

Types of Deeds

The owner of real property is said to hold the title to the property, and a deed is a legal instrument that transfers the title from one party (typically the seller) to another (the buyer). There are several types of deeds, but the two most common types are the *quitclaim deed* and the *warranty deed*.

The quitclaim deed is a simple document that essentially states that one party ceases to have any claim or interest in a title. People often use quitclaim deeds when settling a divorce, an estate, or a gift transfer, but you should never accept a quitclaim deed on real property because it provides you with no warranty. For example, some person of low character could sell you something that he or she didn't even own—say, the Statue of Liberty. Since this person didn't own the property in the first place, you have nothing when you file your quitclaim deed at the courthouse. You might find a prosecutor who'd be willing to charge this person with fraud, but you would have no assurance that you'd ever see a dime of your money again.

When a seller signs property over to you using a warranty deed, on the other hand, he or she is making certain written guarantees that will provide you with some protection against future claims on the property. The guarantees normally include the following:

- The seller has full title and the right to sell the property.
- There are no unspecified encumbrances (for example, liens, mortgages, or easements).
- The buyer is assured the right to "quiet enjoyment" of the land, meaning that no spouse, child, partner, or other party will come out of the woodwork and force you off the land.
- If a problem should arise with the title after transfer, the seller will do whatever is necessary to clear up the problem.
- If another party brings a "rightful claim" against the title, then the seller will reimburse your losses and damages.

If a title has some problems, the seller can go to court to clear the title through a "quiet title" suit. This should be the seller's responsibility. How do you find out about problems prior to purchase? Depending on which state you live in, you'd get either a title search or an update of the abstract, which is prepared through a courthouse records search. If you purchase property with a loan from a commercial lender (mortgage or promissory note), the lender normally requires title insurance. In this case, a professional abstractor or title company will provide a title search as part of the process of supplying the title insurance. If you are purchasing with an *owner-carry loan*, you can do a title search yourself or hire a title company to perform one for you.

A *deed of trust* is a financing system used in some states. In states that use the deed of trust system, a third party acts as a trustee for the financing of property. The trustee may be a bank, a title company, or a government agent (for example, a county clerk). The trustee holds the actual deed until you pay the loan in full. If there is a dispute between the lender and the borrower, the trustee has rules to follow to resolve the conflict. In some states, the borrower pays the trustee, and the trustee pays the financer; in others, the borrower pays the financer directly. If the borrower fails to make payments, it is up to the trustee to dispose of the property and pay off the loan.

A pond can be both beneficial to your livestock and aesthetically pleasing on a larger piece of property.

Land Interests and Codes

Before signing on the bottom line for your dream farm or ranch, there are a few other things you should investigate: access to the land, access through the land, reservations, and zoning and building codes.

- **Access to the land.** Let's say you are looking at a piece of land that isn't adjacent to a public road. There is an old dirt track leading to it through a meadow. The seller or the real estate agent says, "No problem, you can access the land on the dirt track." But, after you buy, the owner

Ponds

According to the US Department of Agriculture (USDA), there are more than 2 million constructed ponds on private lands. Landowners construct them not only for livestock watering and wildlife habitat, but also for irrigation water, fire protection, fish production, recreation, erosion control, and landscaping.

A large pond or one needing a concrete dam will probably require the services of a civil engineer, but, under most circumstances, you should be able to design and construct a small pond without the use of an engineer. If you decide to go it alone, there are many variables you'll have to take into account when selecting a site and sizing your pond, ranging from estimating runoff to assessing soils. In most western states, you should also check with the state's water resources department to find out if there are water laws with which you should be concerned. For example, where I live, in Colorado, it is legal to build a pond that has less than one-third of an acre of surface area on an intermittent stream without a water right, but it is illegal to build any pond along a live stream without owning a water right.

The Natural Resources Conservation Service (NRCS; a branch of the USDA) has an excellent publication on ponds titled *Ponds—Planning, Design, Construction.* It is an eighty-five-page handbook that describes the requirements for building a pond. You can download it from the NRCS website (go to www.nrcs.usda.gov and search for "Agriculture Handbook 590") or order it from your local NRCS office.

of the meadow puts a lock on the gate and tells you that you can't drive through his meadow. Can he do that? The answer is maybe. When a piece of land isn't adjacent to a publicly maintained road, you should require proof in writing that a legal right-of-way exists to access the land. This proof comes in the form of a deeded access easement.

- **Access through the land.** The flip side of this coin is when you look at a piece of land along a public road, and the property has other private, roadless land behind it. You don't give it any thought and buy the property, but one day, several years later, you look out to see a bulldozer driving through the middle of your hay field. *Wait just a minute*, you say, but the owner of the land behind you points out the deeded easement to the parcel he bought behind you. You won't be able to stop this person from tearing up your hay field for his big, gated driveway to his new dream house if he has an easement.

- **Other easements.** Our farm in Minnesota had a utility easement running through the middle of our pasture. There were no utility lines in the ground because the gas company that had acquired the easement decades earlier had ended up taking a different route with the line when they installed it. But they still had an easement and could theoretically show up one day to run a gas line, or they could transfer their right to that easement to some other utility company. Easements aren't necessarily bad, and most utility companies that access their system on your property will work hard to minimize damage to fields and infrastructure (like fences) when they do installations or maintenance. Still, you should know what easements exist before you buy.

- **Reservations.** Deeds can include reservations for certain types of rights, such as mineral, oil, gas, or timber rights. For example, all lands that the federal government offered for homestead after 1901 included a reservation by the government of all mineral rights (including oil and gas). What this translates to is that, at some point, the federal government could authorize a private company to come onto your land to extract valuable minerals. This may never happen, but, then again, new technology may make extraction of heretofore unusable minerals possible. Coal-bed methane wells, which have popped up over large areas of Wyoming, Utah, and Colorado, are a prime example. Although there may be little you can do about reservations, you should be aware of them.

- **Zoning and building codes.** Before buying, you should check with the local government to find out how the land is zoned because this affects its possible uses (and the possible uses of neighboring parcels). You should also find out what kinds of building codes are in effect if you plan to build a new house or outbuildings. Like reservations, there may not be much you can do about local requirements, but if you know what they are ahead of time, you can evaluate their impact on your plans.

Barns, Outbuildings, and Annexes
Barns

Reflecting our agrarian roots, many farms across the country have barns and may sport several outbuildings as well, and your hobby farm may be one of these. The structure of these barns varies widely depending on what part of the country you live in and the background of those who settled the area. Until the early to mid-1900s, most barn designs sprang from traditional knowledge rather than from blueprints. Every builder had his own ethnic and/or regional background, and the local climate, geography, and individual farmer's needs also factored into the barn's ultimate floor plan and construction. For example, in states where winter brings copious amounts of snow, you'll see barns with steep roofs that are designed to let that snow slip off rather than pile up and stress the structure. In the South, however, a steep roof like that would trap the heat more than a shallow design. Too much ventilation in the North leads to chilly animals, whereas too little in the South leads to heat-stressed ones.

All of these elements naturally gave rise to the prevalence of specific architectural types in certain areas. However, even within a region, you will find infinite variations and barns that are hard to classify. Just to throw an extra bit of challenge into the mix, barns are often named after their roof designs rather than their floor plans or construction types. The basic roof styles are as follows:

A gambrel roof (left) compared to a gothic roof.

- **Gable:** Also called a "pitched" roof, this type consists of two parts sloping down from a central ridge, leaving a gable (triangle shape) at each end.
- **Gambrel:** This is, in a way, an adaptation of the gable roof in that it starts out with two parts sloping down from a central ridge. The difference is that those two planes are joined to a second set of planes that drop at a much steeper slope than does the upper set.
- **Gothic:** Although you might hear it referred to as a "domed" roof, a Gothic roof and a domed roof are not actually synonymous, as a true Gothic roof contains a definite peak in the center, from which the roof planes curve down to the eaves. A true domed roof, such as is found on Quonset barns, is indeed a dome, having little or no slope in the middle and no peak.
- **Hipped:** Hipped roofs are similar to gabled roofs, but have no vertical gable ends. Instead, a second set of roof planes angle up from the top wall plate to meet the other two.
- **Saltbox:** Basically, this is a variation on a gabled roof, one plane being shorter than the other.
- **Shed:** Sometimes called a "penthouse" type, this is a roof with one plane.

Although rooflines do often lend their names to barns, there are some classic designs dotting our countryside that are common enough to have their own names.

The English Barn

The English barn, so called because it is reminiscent of barns in England, tends to be small and rectangular—30 feet by 40 feet being the usual dimensions—with a gable roof. Also called a "Yankee" barn, "Connecticut" barn, "three-bay threshing" barn, or "thirty by forty," this style of barn is usually located on level ground and has no basement or loft. Reflecting small-scale, diversified agriculture and the low mechanization of the early colonists, the floor plan typically utilizes three bays that are organized crosswise to the roof ridge—one dedicated to livestock, a central threshing floor, and one for storage of hay or straw. The latter was sometimes used as a granary as well.

The doorway (originally featuring hinged wagon doors) is typically in the center of the north eaves side of the barn and leads onto the threshing floor, with another, often smaller, pedestrian door accessing from the south. One of the first barn styles built in the States, English barns were almost always constructed of timber post-and-beam framing and were popular during colonial times, particularly in the Northeast.

Settlers tended to model their barns after those from their home countries, and most settlers in that area in that era were indeed English.

Transom and wall windows began to show up in the 1800s. Prior to that, ventilation was achieved by air circulating through the narrow gaps in the siding boards. The threshing area had a tight-fitting wooden floor, and, by opening the doors on the north and south sides, the draft would blow away the chaff as farmers winnowed their grain. Typically, this center bay was not at true center

An old-fashioned English barn.

The typical Pennsylvania barn construction, built into a hill and with the characteristic overhang.

because the east bay favored by morning sun was smaller and intended to house cattle. The larger west bay could then accommodate more fodder. Bigger animals—horses and oxen—were sometimes housed on the south end of this west bay. Often, a mow was located above the center bay where farmers could store grain.

The Pennsylvania Barn

The Pennsylvania Historical and Museum Foundation states, "The Pennsylvania barn's main diagnostic feature is the projecting 7–8-foot forebay, or overshoot. The barn is banked on one or both sides (or built into a hill) and organized such that the upper level consists of central threshing floor(s), flanked by mows, and a granary (sometimes in the forebay, sometimes next to a mow on the bank side)." The lower level usually contains stables and stalls for horses, milk cows, beef cattle, and even sheep or hogs. Such compartments are organized crosswise to the roof ridge and separated by alleyways for human access.

These barns, which came on the scene from around 1820–1900, are very versatile in form, accommodating such features as an "outshoot" or "outshed" extending back from the bank side, the aforementioned haymows and threshing floors, a machinery/corn-crib shed extension, and/or a possible machinery bay on the lower level, as well as a root cellar. The jutting forebay might be unsupported or have supporting end walls or posts. Most will have a gable roof and range from 20 to more than 100 feet in length. Ultimately, to be considered Pennsylvania style, a barn must have the essential features of a projecting forebay and banked construction, almost always with the eaves side in the bank/hill.

The short sides and sloping roof of a Dutch barn.

The Dutch Barn

Dutch barns, also called "New World Dutch" barns, have been in America for more than three hundred years. Popular in the 1700s in states such as New York and New Jersey, they can be found all over and are still being built. These versatile structures can be identified by six basic features:

- A square-shaped floor area
- Short sides and a high-sloped roof
- Large Dutch doors in the gable ends (doors that are split in half horizontally so that the bottom half can stay shut while the upper half is open; also called "stable doors")

- Stone piers to raise the floor off the ground
- Smaller side doors near the corners on the gable ends
- "H" framing

Typically, Dutch barns feature a spacious center aisle and a thick plank floor, perfect for unloading wagons, threshing, and other farming needs. Limestone was easily accessible in many parts of the Midwest, and many barns in the Dutch style featured foundations made of limestone blocks, wood-shingled roofs, and rafters and interior framing of hardwood— walnut, oak, maple—cut from the farm. Often, these big barns had three levels: a lower level built into the hillside/bank was used for livestock, a second (ground floor) level for livestock and grain and implement storage, and a third level (loft) was used as a haymow.

In many Pennsylvania Dutch barns, three sides were made with stone, with the south side of wood. Many also had large, elaborate cupolas and air vents, which not only added architectural details but furnished interior ventilation.

The Bank Barn

The bank or sidehill barn is a two-story barn and a common sight throughout the Midwest. Settlers were expanding westward, and, by the 1850s, new, more efficient, and more profitable farming practices engendered changes in their barns. One of these practices was spreading manure on the fields every spring. Sidehill or banked construction allowed farmers to toss manure from the ground level to the lower level for storage, whence it could be moved directly to the fields. They also realized that by housing animals on the lower level, food could be dropped from above, facilitating feeding a larger herd, and the animals could walk directly out into a field/yard. However, it was not recommended to actually house stock in the lower level because ventilation and light were not readily available toward the back regions during winter months.

Traditionally, the upper level provided storage and a threshing floor. One of the beauties of a bank barn is that hilly areas could be utilized as building sites, leaving level ground for such things as gardens while, at the same time, affording a ground-level entrance to both stories (as opposed to other styles that access the second story solely through stairs or ladders). Early bank barns featured gabled roofs, although, later on, gambrel roofs became common.

The Prairie Barn

In the West and Southwest, the prairie barn, sometimes called a "western" barn or "intermountain" barn, was the common choice for farmers because of their need to store a large amount of hay and grain in order to feed during the winter. As settlers brought agriculture further west, prairie barns proliferated during the 1800s and early 1900s. These barns, with their long rooflines and arrangement of animal enclosures on either side of an open central space, are similar to Dutch barns.

The prairie or western barn has long rooflines and a sharply sloping roof peak.

Generally of frame construction with a gambrel or gothic roofline being the norm, foundations tended to be of stone or concrete, with sidewalls of stone for the first 4–5 feet. Tall, framed "western"-style barns tend to have extremely sharp-pitched "saddles" (roof peaks) to help deal with winter snowfall.

Homesteaders used what was readily available to build with, and depending on the part of the country and availability of sawmills, some barns in the West were built of logs, often unchinked, and later sheathed with sawn lumber. You will also see many western/intermountain barns with various-sized shed-roofed wings off both sides, greatly expanding their function and capacity. Barbara Bartell's barn in western Montana's Mission Valley is an example of this design. Built circa 1920, it is not large by some standards, but it has functioned well for more than ninety years. Two shed-roofed annexes—one containing a calf pen with access to an outdoor run/corral as well as to a feed room, and the other functioning as a tack and tool room—expand the barn's function and capacity, while behind the stanchions in the main room, a haymow can be accessed via a built-in ladder. "It was a great barn to grow up with," states one of the Bartells. "When summer nights were warm, we kids used to spread our blankets on the hay in the hayloft and gaze out the big loft window at the stars while purring barn cats lulled us to sleep with their rumbling lullabies." She adds, "Milking the cows morning and night—well, that was not as fun. But it's lovely to look back on, and I feel blessed to have grown up that way."

The Monitor Barn

Monitor barns, also known as "raised-roof" or "raised center aisle" (RCA) barns, have the center portion of the roof raised (or pushed up) from the main roof, supported by short "knee" walls. The look of these barns can be altered by making that center part higher, lower, wider, or narrower, giving them a unique look that many prefer over a regular gabled roof. But the purpose of the raised roof was more than just to add architectural interest. As noted with the lower level of the bank barns, darkness and lack of ventilation leads to health issues for livestock. Raising the center portion of the roof and installing ventilation shafts from the ground level to the openings in the center section addressed this necessity, allowing constant air exchange, even in winter. The addition of cupolas added further ventilation and architectural interest.

Stalls/pens for animals usually occupy either side of the main center floor. Often, horse barns follow a monitor design, with stalls opening into the center aisle, although this floor plan is not unique to this particular architectural style. Monitor/RCA barns tend to be more expensive to build than their gable-roofed counterparts.

The Round Barn

The round barn design is perhaps the most unique type seen in the United States. They have a huge haymow capacity, with a loft-floor diameter of as much as 80 feet. Advocates of the style in the early days believed that the round shape saved them construction costs and offered more efficient space than did traditional right-angle-constructed barns. In a Kansas State Board of Agriculture biennial report dated 1911–12, the author writes, "When a country man is convinced that he can save from thirty-four to fifty-eight per cent of the cost of a rectangular barn by constructing a round barn of quite similar area, he usually becomes decidedly enthusiastic about this unique building, other things being equal."

The unique and not-often-seen design of a round barn.

While savings in our day and age would probably not approach that of early 1900s prices, the beauty and functionality of a round barn still holds attraction. A round or octagonal shape topped by a self-supporting domed roof yields an amazing amount of interior space, which farmers used to house dairy herds, horses, and draft animals. Some even had built-in overhead conveyor tracks on the perimeter of the lower level, which were used to move equipment for feeding and manure removal.

Another unique feature of some round barns is a central shaft for ventilation that was also used to drop hay and other feedstuffs from the loft, while a certain number of round barns featured dumbwaiter-type elevators. Others had equipment to grind feed installed at the bottom of the shaft.

Barn Functions

As is true for most architecture, form is married to function. Although certain barn styles do seem to be synonymous with specific parts of the country, giving witness to their historic agrarian past, the reality is that the floor plan and how a particular barn functions are more important than its exterior features, no matter where you live. Barn functions fall into three basic categories—housing and working of livestock; storage for hay, feed, farm implements, and machinery; and space to work and repair machinery—with multiple adaptations and combinations thereof.

If you're looking at building a barn, ask yourself *why* and *what*. Why do I need a barn? What will a barn offer that I do not currently have? These types of questions will help you define what components you want to

Solar Barn

Whether you're retrofitting your existing barn or planning a new one, you might just look into solar power. Solar power may be more expensive to install initially, but, over the lifetime of your barn, it can pay for itself, and if your ideal barn location is far from a power pole, it may not turn out to cost much more than running power lines over long distances. Solar energy can heat both water and your barn during cold winter weather without massive heating bills. Solar companies can assist you in assessing your site for suitability and, if solar-friendly, designing a system that works for you. With enough sun and a big roof, you might even end up selling power to the electric company!

include in your barn. No matter your ultimate plan, all barns should have these key components:

- **Ventilation:** Introduce as much ventilation as possible through the use of vents, louvers, cupolas, open ceilings, windows, and sliding or Dutch doors opening into paddocks or corrals. Barns can be fitted with fans to increase air flow as well when the temperatures are high.
- **Lighting:** Provide natural lighting via windows, skylights, and open doors. Artificial lighting is beneficial as well, especially when you need to make night visits to the barn to check on ailing animals or those about to give birth. (Even if you don't plan on housing animals in your barn, keep resale value in mind.)
- **Safe wiring:** House all electrical wiring in rodent-proof metal or plastic conduit and out of reach of animals, such as horses, who like to chew on things. Give yourself plenty of electrical outlets to minimize the need to string extension cords hither and yon. To reduce fire danger, situate light fixtures so that they cannot be accessed by animals or children.
- **Well-thought-out floor plan:** Need more be said? Whether you're remodeling or starting from scratch, the time to plan is now. If you'll be using your barn for several purposes, define those purposes and how they will overlap/interact. Although the scope of this chapter doesn't allow for an extensive discussion of floor plans, it might be helpful to know that resources for barn planning abound all over the country, as do architectural standards, blueprints, history, and barn-building companies. You'll never regret time spent designing for optimal efficiency and function.

Build horse stalls according to recommended specifications and your horse's size.

Livestock

If you find that your existing barn requires some adaptation to meet your livestock's needs, or you are about to embark upon a barn-building adventure, your first step should be to research optimal design features via the Internet, the library, your neighbors, and your extension office. Milking parlors have standard recommended measurements, as do horse stalls and indoor pens for calves or other young stock. In general, horses are about the only animals that are routinely housed, and they (as well as most other livestock) are generally quite content with a three-sided shelter called a "loafing shed," except in the harshest of weather.

Storage

Storing hay and straw can take up a whole lot of space, and many barns have haylofts/mows for that very purpose. Just make sure that your barn can withstand the weight of several tons if you do indeed store that much forage up there. Barns also often have a feed room where grain can be stored. Although it

An amply sized barn can serve multiple purposes, including storage for hay and equipment.

may be impossible to completely eradicate mice, field rats, mealy bugs, and more, make your feed room as pest-free as possible with bins that resists gnawing teeth and keep moisture at bay and a door to exclude larger scavengers like raccoons, possums, skunks, woodchucks, all of which may be drawn to the smell of your grain-based feed (especially if molasses is in the mix). A couple of barn cats will go a long way toward keeping the rodent population in check, not only in your barn but also in your fields. Install a latch on your feed-room door that resists children and animal entrance to minimize chances of the door being left open. A cow or horse with access to an accidentally open feed room can overeat and potentially bloat.

As I mentioned previously, in the old days, farmers stored their wagons, harnesses, buggies—whatever they needed to—in their barns, and although wagons have given way to modern counterparts, barns are great places to park small tractors and other equipment like rototillers and wood splitters, thus keeping them out of the weather.

Workspace

The center space in a barn, that which is not occupied by animals or storage, was once traditionally a threshing floor during harvest season but nowadays makes a grand place in which to build projects, repair equipment, groom and saddle a horse, or—my favorites—play with the latest batch of barn kittens or just sit and enjoy the unique atmosphere that a big old barn provides.

Location

Although early farmers sometimes located their barns away from their houses, by the 1850s, farming journals were advocating that barns be built close to the house to consolidate efforts to a smaller area. Some farmers even went to the extent of connecting them to their houses!

Obviously, this is probably not an optimal choice—can you imagine the fly problem in the middle of summer, to say nothing of odor!—but it's good to keep in mind that well-placed barns can serve as windbreaks for your house and yard. If you're dealing with an existing barn, it is where it is, even if you wish it occupied some other spot. But for those who are contemplating building, ponder location. It is huge. Here are some things to keep in mind as you seek the perfect building site.

Drainage

One horse rancher, Christine Churchill at Colorado's Five Star Ranch, wrote that one of the smartest things she did was to walk her proposed barn site during a heavy rainstorm. This revealed the natural drainage. Although it is possible to create good drainage by steering water away from your site via landscaping and drainage tiles, obviously, the more modification you have to make to your site, the more your barn will cost. She also suggests avoiding low places where moisture gathers and insects breed. If that's all you've got, bringing in fill may be your only viable option. Don't skimp! High and dry is the aim. You certainly don't want your barn flooding every time the weather clouds up. Google Maps can be a good starting place for observing land contours, but walking the property during or after a week of rain will really narrow down your choices for barn sites. This is good. If you've got a hill, so much the better, but do consider setting the barn below the crest of the hill to help shield it from direct winds. Unless you're building a bank-type barn, an ideal site will be on level ground. Again, this may take some excavating or filling if no such spot occurs naturally.

Convenience

Having your barn(s) convenient to driveway and house just makes sense. After all, unless you just want to ensure that you get plenty of exercise walking to and fro, you don't want it in the north forty. That said, neither do you want it so close that every time you open a window in your house, you smell the barn. A happy medium would be to be able to see the barn (with your animals happily loafing nearby or with their heads hanging out of their stalls) yet not smell it. Olfactory concerns aside, you'll appreciate proximity if you need to make midnight trips to the barn to check on the livestock, and having it accessible via a driveway becomes extra important during the height of monsoon season or when your vet needs to make an emergency visit. Vets often carry mini-clinics in the back of their vehicles, and being able to drive right up to the barn becomes imperative under emergency situations. A solid drive and parking area is also awesome

A conveniently located barn makes caring for and protecting your animals easier.

when it comes to delivering a load of hay, loading and unloading animals, and bringing tractors and equipment in and out of the barn/barnyard. The good (although perhaps a bit pricey) news is that you can always add a graveled drive to access an existing barn if it doesn't have one.

Proximity

Locating your barn near to your house can also help cut down on trespassing, which in some parts of the country has become a real issue. Yes, you can lock your tack room (although if thieves have enough time, a mere

A barn that provides shelter for livestock can open into a grazing area.

locked door won't deter them much). But people whose barns set too far away from their residences have fallen prey to livestock and trailer thieves, and with copper prices on the rise, even the wiring can be at risk. And speaking of wiring, the closer you can build your barn to an existing power pole, the cheaper it will be to connect to a main power source. Old-time farmers may have had barns without lights, but it is not as romantic as it sounds. The aforementioned Missouri barn circa 1885 had been retrofitted with an ancient breaker box that boasted two electrical receptacles (from which extension cords snaked to power brooder lights and water tank heaters) and two very small light fixtures hanging high up near the peak of the 30-plus-foot roof. Next to no help at all in the dark of night when investigating a panicked flock of laying hens housed in one of old corn cribs within the structure! No matter how bright they are, flashlights just don't do an adequate job in the recesses of an old barn when you're not sure if those glowing eyes are from the resident barn cat or a marauding raccoon, possum, or skunk. No. You will want electricity in your barn, even if you choose not to use it regularly.

Proximity to water lines is something to consider as well. Even if you don't install plumbing in your barn (although if you can, you will not be sorry!), you'll need water for your animals, so if you're deciding between several sites that are equal in other respects, choose the one closest to power and water sources. Another consideration: If you've got other outbuildings—hay barns, loafing sheds, granary, storage shed, corrals, round pen, arena—you'll want the barn close enough to these that hay and grain can be accessed easily, and it's a short walk from barn to destination. And finally—although the location of your manure pile may not be top of your list of items to be aware of, in reality, you'll want to plan for this, either because you'll be spreading it on your own fields or gardens or offering it to friends. Churchill writes, "Obviously you don't want it in full view, you want it out back, so make sure there is a place that is convenient for your manure pile. I like to make sure there is truck access to the pile because I have a number of friends who will come and take manure regularly if I make it convenient. I've done this in several barns now and never had a large manure build-up. Make it easy and they will come."

A brick-red barn complemented by the rich hues of fall foliage.

The View

Another factor you'll want to consider when locating a barn site is whether or not it is connected to pasture so that stock can come and go without your having to lead or herd them. And last, although certainly not least important—you may like to look out at your barn, but chances are that you don't want it blocking your favorite view! Churchill took several pictures of their back pasture from the vantage point they would usually have and then used a photo editor to join these pictures together, giving a panorama of possible barn sites. Next, they dropped in a scaled picture of the proposed barn structure, moving it around on picture to see how the barn would look in different places on the property until they found one that wouldn't ruin their view.

Direction and Climate

Although we all know that different parts of the country have their own climate, it's easy to forget this detail if you've relocated. Churchill points out that people tend to build the type of barn that they have been exposed to. For example, someone from a colder northern region where barns are built as airtight as possible can forget that a barn in the southern parts of the country needs big-time ventilation. Not only the structure itself, but the positioning of the barn is important. Situating your barn to take advantage of seasonal sunshine, shade, and breezes will aid in temperature control. She writes, "When I built a barn in Kansas and Colorado, I built for warmth. I wanted as much sun exposure as I could get, so I positioned my barn facing south to take advantage of solar heating and designed a barn that had a solid wall against the prevailing winter wind. In hot Texas, I had to design my barn and site selection completely differently. There, I avoided facing the stalls to the west because of the hot afternoon sun, and I tried to position the barn to capture prevailing breezes. With my Texas barns, I always try to include 12-foot porches so the horses can stand in the shade but catch a breeze." She adds that her farrier loves the porch, too—better light and a nice breeze.

Finishing Touches

We've all seen pictures of old red barns against a backdrop of flaming hardwoods in the fall or weathered gray barns silhouetted against the jagged Wyoming mountains. Although it may seem like a small detail at the time, you won't regret selecting a barn design and color that complement the countryside. In addition, landscaping can help integrate your barn into its surroundings, maintaining and even enhancing your view, which adds to your property value as well as your personal enjoyment. While you're in the process, keep in mind that at some point you may want to expand your facilities—add an arena, annex a loafing shed, and the like. So pick your location with potential expansion in mind.

Let your barn's design reflect your creativity and personality.

Annexes and Outbuildings

As most farmers discover, sometimes one barn just isn't sufficient. You need more space to store hay or grain. You decide to add beef or dairy cattle to your menagerie. Or chickens. Or goats, pigs, or some other four-legged barnyard denizens. If your existing barn does not provide enough room or lend itself to the venture, consider adding an annex or outbuilding to your hobby farm.

Annexes

By definition, an annex is something attached, added, or appended to a larger building. Keeping in mind the overall aesthetic balance of your barn, sometimes adding on is the most cost- and labor-efficient choice, and an annex has the advantage of already available power and water. Typical annexes tend to utilize shed/lean-to roofs, attached along the barn's existing dripline. When the owners added a loafing shed to the side of their 1885 Missouri barn, they used pole barn techniques, used crushed limestone to give good drainage and a firm floor, and adapted the swinging door that led out the south side into a sliding door that wouldn't collide with the roof of the lean-to. Especially in the winter, this annex, open all along its length to the south, became a favorite hangout for the family's Jersey milk cow, Rosie, and her sidekick, an aged Welsh pony named Ginger.

Although the owners used new heavy-grade tin for the roof, they utilized old tin that matched the current barn siding for the ends, so that unless you were in an airplane flying over, the annex appeared to be part of the original barn. These particular hobby farmers added chicken runs on the opposite side of the barn and housed their chickens in what had originally been granary rooms inside the barn. By cutting a small door in the side and giving the chickens a ramp to access the covered run, their chicken operation was up and running in no time. However, if the barn had been configured differently inside, an annex matching the one on the south side would have been a good option as well. Some barn annexes, like the Bartell barn mentioned earlier, are not open on one side

but are completely sided to match the barn.

If you plan to store hay in your annex, don't forget a door large enough to accommodate your pickup, trailer, or whatever you use to transport hay. Farmers who have transitioned from small bales to large round or square bales will need to allow room for a tractor with which to lift, position, and remove hay as needed. Other possible uses for a barn annex are housing livestock, machinery, or tack; office, storage, or tool shed; machine or woodworking shop—pretty much whatever you need it to be!

This barn in northern Idaho features a grain silo from its original construction decades ago.

Outbuildings

Annexes are great, but sometimes you just need more—more hay storage, more room for your chickens, more space to store tractors and equipment. This is why, when you drive through the countryside, you typically see farms with a house, a main barn, and several outbuildings of various sizes and shapes. There is no "right" use or number of these that make a farm a real farm. Need and function are the determining factors, and only you know what you want to do on your farm. That said, remember that your farming neighbors can be a great source of experience and ideas. A great question to ask them would be, "If you could design an ideal chicken house (loafing shed, tractor storage, etc.) what elements would you include?" The internet, of course, is full of information. Just make sure you've accessed a reputable site that understands function and practicality as well as aesthetics. Check out your local extension office as well because they generally have booklets on all things farm-related.

Common outbuildings are:

- **Chicken coop:** With or without an attached run and including nest boxes and roost(s) (for more information, see chapter 7). You will want to wire it for electricity so you can plug in brooder lights or a heated waterer for cold winter months.
- **Loafing shed:** Also called a "stall barn" or "pen barn", beloved by cattle, sheep, goats, horses, and the like. This is a place for animals to get out of the sun, rain, wind, and snow. Some loafing sheds are merely a roof on poles, perhaps with wooden fence panels on three sides. Others have walls on three sides, with or without fencing and gate(s) on the front. Again, what you need for your particular operation will help you settle on a design.

- **Pigpen:** If you want to raise pigs, they'll need their own pen—usually a three-sided, roofed building, like a small loafing shed, with an extended, fenced yard/pen where they can create their mud wallows (that protect their skin from sunburn and biting insects), root around, and be happy. For your own happiness, locate a pigpen downwind of the prevailing weather so you can open your house windows and not be treated to less-than-desirable odors.

- **Hayshed:** This can be merely a pole barn without sides with a fence around it, a fully enclosed version, or some variation thereof. Farmers who use small bales have been known to build in feeders along the sides so that they can feed hay directly to stock. Haysheds often do not have electricity, but if you can install the basics when you build, retrofitting becomes much easier and less expensive if you want to wire it at some future time.
- **Machine shed:** Although bigger doesn't equate better, be sure you don't build too small. You might want to upsize your equipment at some point or add another tractor or baler, so, as much as possible, build with your future in mind. You'll want electricity because, typically, not only do farmers store their machinery in these sheds, they also perform routine maintenance and repairs. Floors are usually gravel or concrete.
- **Granary:** Although storing grain is not as common on small farms as it once was, many farms still retain a granary or two from earlier times. On the Bartells' Montana farmstead, the granary is of wood and built off the ground to raise the floor above moisture level during winter and discourage burrowing varmints like skunks from taking up residence below it. Currently, it stores odds and ends instead of fodder. Other farms might have the classic round metal granaries with raised concrete floors, again to keep moisture at bay.
- **Garage:** Self-explanatory. Many farmhouses don't have attached garages, and farmers need garages just like everyone else.
- **Wellhouse:** Many farms have little outbuildings to house their wells and pressure tanks. Chances are if you need one, you've already got one.
- **Outhouse:** Outhouses (outside toilets), also known as "privies," were standard on early farms but are increasingly rare, mostly due to the fact that they are no longer as simple as digging a hole and putting up a little house with one or two "holes." However, privies/pit toilets/earth toilets/whatever-they-term-them-in-your-part-of-the-country are still legal, but highly regulated, so be sure and check your state and local laws before digging. Your county sanitation department may be a good place to find pertinent information on location and construction requirements.
- **Storage sheds:** You can't have too much storage. Beware, though—it does seem that the more storage you have, the more stuff you acquire to store!

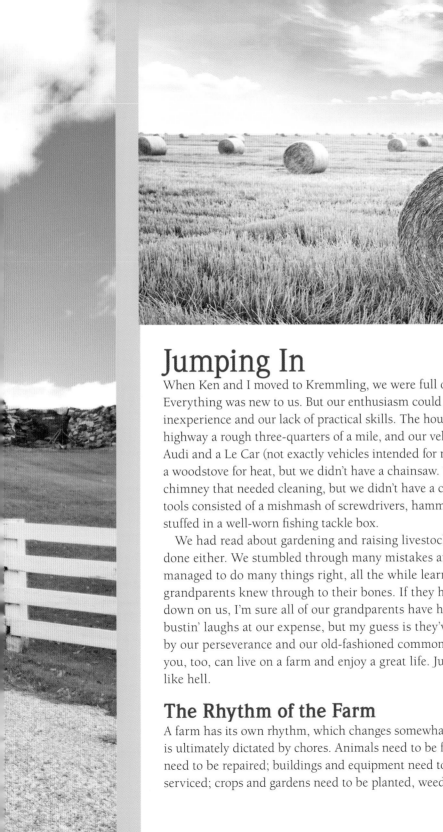

Jumping In

When Ken and I moved to Kremmling, we were full of enthusiasm. Everything was new to us. But our enthusiasm could not make up for our inexperience and our lack of practical skills. The house sat back from the highway a rough three-quarters of a mile, and our vehicles consisted of an Audi and a Le Car (not exactly vehicles intended for rough terrain). We had a woodstove for heat, but we didn't have a chainsaw. The woodstove had a chimney that needed cleaning, but we didn't have a chimney brush. Our tools consisted of a mishmash of screwdrivers, hammers, and wrenches stuffed in a well-worn fishing tackle box.

We had read about gardening and raising livestock but had never done either. We stumbled through many mistakes and problems, yet we managed to do many things right, all the while learning skills that our grandparents knew through to their bones. If they have been looking down on us, I'm sure all of our grandparents have had some good belly-bustin' laughs at our expense, but my guess is they've also been impressed by our perseverance and our old-fashioned common sense. I know that you, too, can live on a farm and enjoy a great life. Just jump in and paddle like hell.

The Rhythm of the Farm

A farm has its own rhythm, which changes somewhat by seasons but which is ultimately dictated by chores. Animals need to be fed and watered; fences need to be repaired; buildings and equipment need to be maintained and serviced; crops and gardens need to be planted, weeded, irrigated, and

Even when the fields are covered with snow, there is work to do around the farm.

harvested. The harvest needs to be put up, stored, and possibly marketed. Let me run down our current chores by season.

Winter

Winter is the time to plan and study, the time to do as much indoor work as possible. We burn wood as our primary heat source, so firewood dominates the outside chore load at this time of year. Ken fells trees that have succumbed to old age, insects, or disease. He cuts and splits these into pieces that fit in our stove. I actually love to split wood, and I help regularly with this chore.

The chimney requires regular cleaning to reduce the likelihood of an accidental chimney fire (which can easily lead to a house fire) and to keep the fire burning well. We have found that cleaning the chimney about four times a year is sufficient, but the optimal number of annual cleanings can vary depending on the type of wood you burn and the kind of stove you burn it in, the length and type of your chimney, and the strength of the draw. Airtight stoves and stoves with catalytic converters tend to reduce the creosote buildup in the chimney because they burn more efficiently. The 1885 farmhouse in Missouri boasts 30-foot chimneys, relined and fitted with airtight stoves that burn from September to April as the main heat source yet leave such a thin layer of residue that the chimneys need cleaning only once every other year. If you don't want to do this chore yourself, you can call a local company to clean your chimney and assess your stove's safety and efficiency.

This time of year also makes for more onerous animal chores. The animals are not getting much feed from grazing (or, if the snow is deep, no feed at all), so we feed hay morning and night. We also provide fresh water daily. Our watering system requires us to run about a hundred feet of hose, which has to be well drained each day to protect it from freezing. Sometimes the hose freezes despite our best efforts, and we must drag it into the house to let it thaw. Floating tank heaters can keep your water troughs ice-free, once you get them filled.

For smaller critters like chickens, bringing hot water from the house to thaw and fill water pans is a time-honored method to weather the winter months, although other means are available these days. Some farms have automatic stock waterers with buried water lines that are much less prone to freezing. But if your farm isn't one of those, you can explore the multitude of options available to winterize your animals' water. As many farmers before you will testify, whatever eases the workload is welcome!

Doing outdoor chores when temperatures are falling well below freezing (here, we have had temperatures as cold as –37 degrees Fahrenheit; in Minnesota, we had –52 degrees) can sap your energy. Trudging around with twenty pounds of extra clothing—winter overalls, heavy coats, big boots, hats, gloves, overmittens, and face protection—adds to the work, as does having to break trail when there is new snow on the ground. These are the times when you can lose touch with why moving to the farm was such a brilliant idea, and you'll understand a bit more why old-timers seem so much less excitable about the whole experience.

When winter grips your ground, and spring seems far away, something fresh and green can be mighty welcome. I maintain indoor planters full of greens during the winter, reseeding them on a regular basis to maintain a continuous supply. I have found that EarthBox-type planters work well for this application in the winter; I then use them outdoors in the summer.

Spring

Spring is glorious and busy, although it is also known in these parts as "mud season." Early spring—which is really just the unexpected nice days when winter begins to relinquish its hold on the earth—is a time for cleaning up around the place, digging garden beds, checking on the fences, and performing minor repairs that had been on hold, waiting for a fine day. As early as possible in the spring, I try to turn the winter accumulation of compost. I start only a small number of plants inside these days because our house is tiny, but when we lived in a bigger house, I lined every window with soil blocks and plants to transplant later in spring or early summer.

These are also the days when the animals begin to shed, so we often spend time currying them to get out the excess winter hair. They love it, and so do we. This is a fun time for simply watching the critters. They play wildly and with abandon on these early spring days. Our donkeys, Jessie and Duke, will have mock stallion battles that rage for hours. Wild things begin moving around more if they are year-round residents, and the "snowbirds" begin returning, first in ones and twos, and then in droves. We are thrilled to see the first mountain bluebirds return, flashing a streak of intense blue past a window in early March.

The animals enjoy the warmer weather and greener pastures of springtime.

The farm in summer is in full bloom.

Late spring is the time of green. I begin planting outside in the garden, and Ken starts getting construction projects going (there seems to be at least one every year). It is also the time when colts, calves, and lambs are hitting the ground running, and chicks arrive (via the US mail). Life is exploding everywhere.

Summer

Summer is the season for a lot of work but is also a time for enjoying the fruits of our labor. The garden begins providing some fresh greens and other short-season crops and will continue to produce through late fall. Even a small garden provides an unbelievable bounty. As time allows, I try to can some of the harvest (as well as some produce purchased at farmers' markets) for later use.

Construction, repairs, and maintenance are now in full swing. The work always takes longer than we had hoped, but we have come to understand that this is the way of do-it-yourself projects.

During the summer, we try to allocate some time for fun things, too, such as riding horses, hiking, and fishing. Ken brings home fresh trout often, and we fry it the same evening with a touch of olive oil in cast-iron frying pans. Our only chance of traveling is during the summer because the must-do-daily workload is at its lowest of the year. Sometimes Ken's dad comes up and stays so we can get away for a few days.

Fall

Fall is a splendid season. The pesky flies and mosquitoes disappear as the nights get colder. The aspens blaze against the blue sky in yellow and gold exaltation. Birds are drunk on the fermenting currants. Around here, the roads are almost empty because summer tourists have left, and skiers have yet to show up. We enjoy the crisp days, spending as much time outdoors as possible.

In autumn, we harvest the remaining garden produce (the root crops are ready for storage) and clean up the garden. It's also hunting season, and Ken regularly participates in this food-gathering tradition—something his grandfather, father, and uncles taught him as a boy. He hunts with a muzzleloader, a time-honored weapon that is loaded with powder and ball, a weapon that had its heyday in the early nineteenth century. A muzzleloader holds only one shot, which you must shoot at close range. Ken doesn't succeed in shooting an animal every year, but when he does, we eat elk, deer, or antelope—healthy, lean meat—through much of the winter.

Because we no longer have hay fields, we buy all of our hay; we try to get a large portion of our year's supply purchased and stored now. It is the time to start collecting firewood again, too. We try to have at least half of our winter's wood supply before fall gives way to the deep freeze. Some folks lay in their supply in early summer to allow for the wood to dry because wet wood creates more creosote buildup in the chimney. As night temperatures cool off enough to send us to the stove for a small fire, the chimney gets its first cleaning of the season.

Accessing Experts

Old Abe Lincoln may be best known for his role in ending slavery in this country, but he also performed an important service that still benefits farmers and ranchers each and every day. In 1862, Abe signed the Morrill Act into law, which provided states with at least 90,000 acres of land (the actual amount was tied to the number of representatives the state had in Congress) that the states were to sell, with the proceeds earmarked for

Harvest time in autumn includes pumpkins, squashes, and gourds.

the endowment of "land-grant" colleges. These land-grant colleges were to "teach such branches of learning as are related to agriculture and mechanic arts."

In 1914, President Woodrow Wilson signed another act into law. The Smith-Lever Cooperative Extension Act authorized federal support at the state level for extension services associated with each US land-grant college. Today, land-grant colleges not only provide agricultural education for students but also play a key role in research. Most US counties have a cooperative extension office, with extension agents to help disseminate information and act as a bridge between the university researchers and farmers.

Your local extension agent is one of the first people you should meet. He or she is there to help. Although no one person can know the answer to every question or have a solution to every problem that might crop up, extension personnel have a weblike support structure that allows your agent to find an expert who can answer your question or help solve your problem if he or she can't. Extension agents have a wealth of free or low-cost literature on subjects ranging from apples to zucchini, air conditioners to wells. They can also help interpret information and clarify concepts that may be new to you, such as deciphering a soil report from a lab and determining an approach to improving soil health.

The extension service also runs 4-H programs and provides a backbone for county fairs throughout the country. Through 4-H, kids develop confidence at the same time as they learn new skills. Programs are still largely centered on agriculture, animals, and home economics, but there are also opportunities for kids to learn about construction, photography, music, rocket science, and just about anything else that might interest them and that has a practical side. I participated in 4-H from the time I was nine until I was eighteen, and I believe it was the one of the very best things I ever did. It obviously hasn't become a thing of the past because both Michelle Huerta and Judy Cavagnetto say that their kids have enjoyed and benefited from participation in 4-H. Local chapters thrive all across the country.

Most counties also have offices (or have access to offices in a neighboring county) of the Natural Resources Conservation Service (NRCS) and the Farm Service Agency (FSA). These are both

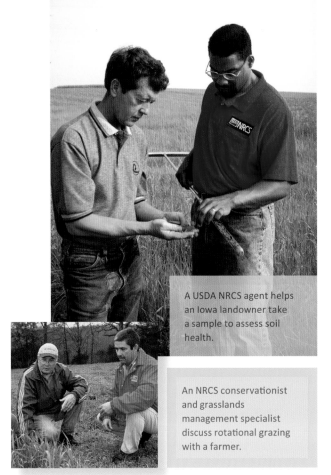

A USDA NRCS agent helps an Iowa landowner take a sample to assess soil health.

An NRCS conservationist and grasslands management specialist discuss rotational grazing with a farmer.

agencies of the US Department of Agriculture (USDA) that were started during the later years of the Dust Bowl and the Depression. Today, the NRCS is the lead federal agency in the effort to help people conserve, maintain, and improve natural resources and the environment on private lands, working with farmers and other landowners. You can have pastures, ponds, and more assessed by NRCS agents, and experts are available at these offices to make recommendations for improvements and advise you on conservation programs and projects. Many even have cost-share funding available for project implementation.

The FSA acts as the financial arm of the federal government with respect to agriculture. It works to stabilize farm income, help farmers conserve land and water resources through administration of certain funding programs, provide credit to new or disadvantaged farmers and ranchers, and help farm operations recover from the effects of disaster.

Shop around for a vet. Large-animal veterinarians who come right to the farm with their mobile clinics are generally available in areas where there is a tradition of livestock farming or hobby horse operations, but they are not as plentiful as they used to be. I'd advise developing a professional relationship with your veterinarian before you actually need to have him or her out for an emergency. Be sure to ascertain if the vet does indeed make farm calls and whether he or she is available on evenings and weekends because somehow animals don't seem to choose the best times to get sick, become injured, or have trouble delivering their young. A little precrisis homework will stand you in good stead. It's one of the most helpless feelings ever to have an animal in critical condition and to find that the first five vets you call aren't interested in interrupting their weekend or evening to come to your rescue. Ask your farming neighbors which vet they call, what they like about him or her, and how easy the vet is to get hold of in emergencies and in situations where transport is not possible. Good vets don't only doctor the critters; they also act as advisers, answering your questions and making suggestions to improve both your animals' health and your bottom line.

Don't overlook the experts in your own backyard—your neighbors. Not only can they recommend a good vet or someone to trim your horse's hooves, they willingly answer question upon question about farming and farm life in general. You should listen to their nuanced, firsthand wisdom, which exceeds all of the book knowledge in the world. At the same time, glean all you can from research,

Beginning Farmer Loans

One FSA program worth knowing about is the Beginning Farmer and Rancher Development Program for new farmers and ranchers. The loan program helps beginning farmers and ranchers with low-interest loans. For purposes of the program, a beginner is an individual who has not operated an agricultural enterprise for more than ten years; who materially and substantially participates in the farm or ranch; who provides substantial day-to-day labor and management on the farm or ranch; who demonstrates a need for assistance; and, for the purpose of farm real estate loans, who does not own land that in aggregate exceeds 30 percent of country average farm size. This definition applies to all credit programs and to conservation cost-share programs. Go to the FSA's website (www.fsa.usda.gov) for more information on the program or to find your state's FSA office.

extension agents, and publications, and don't be afraid to try new things that local farmers may not be implementing. An adage that most farmers relate to is "if it's not broke, don't fix it." If what they're doing is working for them, there's little reason to change, even when they're aware that newer options are out there. However, if your use of a new product or method yields superior results, giving true return for your investment, a neighbor may decide to update his or her own operation in similar form. Farmers are practical.

When we first moved to the farm, no one in our immediate vicinity was doing managed grazing or using electroplastic wire (a thin and lightweight fencing wire made of plastic twine with tiny strands of aluminum woven in to carry the current from an electric fence charger) for fencing. When we showed our neighbor Willy our two-strand fence around a previously unfenced cornfield that we had converted to pasture, he fingered the plastic wire strung on plastic fence posts that are simply "stepped in" to the ground on short metal spikes and said, "Well, that's different." I'm sure he expected to find our cows in his cornfield come morning, but I'm thankful to say that he didn't.

If you propound ideas such as organic production, niche marketing, specialty crops, or agritourism, don't be surprised if some of your neighbors aren't quick to adopt these trends. But, at the same time, they might begin asking you how you did something because, after all, they too want to improve their holdings and their earnings just like you do. For example, we became involved in the Sustainable Farming Association of Minnesota at its inception and hosted one of its earliest field days in the central part of the state. A number of traditional farmers from right around our area came and asked questions. By the time we moved back to Colorado, many had adopted ideas presented through such events, including the aforementioned easy-to-install electroplastic fencing for subdividing paddocks in a rotational grazing system.

A valuable resource may be as close as next door.

Essential Tools of the Trade

Running a farm requires a variety of tools and equipment; some are highly specialized, others rather general. Over the course of years, you will accumulate a variety of implements, but, to me, this is the list of items you should acquire first and as soon as possible.

Vehicles

Pickups: Before we bought a pickup, we bought an old Chevy Blazer. We quickly realized that an SUV is not, and never will be, a pickup. A pickup is one of the most versatile and useful long-term investments you'll make for your farm. It is indispensable for hauling everything from building supplies to feed, firewood, fencing, and sod. When you team a pickup with a trailer, you can move livestock and much more. Although a pickup is a big investment, it is an investment that can last for decades when well cared for (remember to change that oil every 3,500 miles!). You can often find a sturdy used pickup for a reasonable price.

Toppers: Add a basic topper to your pickup bed, and you can keep items dry while hauling, use the truck for camping, or transport small animals. We've gotten some amused looks at gas stations and other stops over the years when the topper held a menagerie. We once drove thirty rather large turkeys to the poultry processing plant; another time, we had half a dozen ducks and geese and several big rabbits hanging out in the back. You can also easily move goats, sheep, and small pigs under a topper.

Tractors and ATVs: Even if you dream of working your land with animals, a tractor or a farm utility vehicle (an all-terrain vehicle [ATV] or a cargo-ATV, often called a "ute") is a practical thing to have; they are available through tractor dealers. Although we like to use our donkeys for packing firewood when time permits, winter often finds us out with the ATV, equipped with its own trailer, bringing in wood. Compact tractors are versatile pieces of equipment that are perfect for small farms. Depending on the attachments you have, they can do just about anything. Mowing acres of lawn, digging a foundation, baling hay, pushing snow, setting fence posts, brushhogging: it's all possible with a compact tractor.

A pickup truck doesn't have to be fancy to get the job done on the farm.

For those in the under-60-acre category, a ute will often do all the work you need done. Manufacturers now make attachments ranging from front-end loaders to manure spreaders that are specially designed to work with these rigs. The fat tires, light weight, and low center of gravity make these units easy on the land.

If money is not a big issue, and you have a larger piece of land, having both a tractor and a farm utility vehicle is the way to go. The tractor can do heavy work, such as baling hay, but the utility vehicle can haul supplies or get you across the farm quickly to move a paddock fence or check on the stock and crops.

Two-wheel-drive trucks, tractors, and utility vehicles are OK for flat ground that's not too muddy, but if you're on hilly terrain, in a snowy climate, or frequently in muddy conditions, invest the extra money up front in four-wheel drive.

Stock trailers: If you plan to have large livestock (horses, cows, llamas, yaks), consider getting a stock trailer. We don't actually use our trailer all that frequently for hauling livestock, but I still consider it one of the best investments we've made in twenty-some years. When we do need to haul our animals, we don't have to borrow or rent a trailer. In between animal trips, we have used the stock trailer to haul firewood, feed, hay, and lumber. During the several household moves we've made since we purchased our trailer, it has done duty as a moving van, hauling furniture and household items. In addition, it often serves as a temporary storage shed. Although two-horse trailers are readily available, opt for a bigger trailer that can accommodate four or more animals. Those two-horse trailers may look cute, but horses ride better in the bigger trailers, anyway.

A shiny new four-wheel-drive tractor.

Smaller Equipment

Handcarts and wheelbarrows: I suggest investing in a handcart right away. Priced between $100 and $400, depending on capacity, handcarts are bargains for how useful they are. They are lightweight, yet they carry big loads easily. Unlike their single-wheeled cousin, the wheelbarrow, handcarts are sturdy and steady over uneven ground or through mud or snow. And they are just plain hard to tip. Best of all, they are balanced in such a way as to be easy on the back. Some carts come with an accessory that converts them to a trailer for pulling behind a ute or small tractor.

We do have a two-wheeled wheelbarrow, which is more stable than the one-wheeled version, and we use it and handcarts for everything around the farm. In the garden or yard, they're great for moving topsoil, plants, seeds, fertilizer, and tools. In the barn, they are handy for cleaning up small piles of manure or dragging tack from point A to point B. They are a perfect size for moving a bale of hay or a few fifty-pound bags of feed. We have also brought newborn calves from the field to the barn in bad weather, with mama tagging along right next to the cart.

Manure spreaders: If you are going to have any livestock, sooner or later you will need a manure spreader. Even if your animals will spend most of their time on pasture, piles of manure accumulate. Spreading this manure thinly over the land improves fertility, reduces contaminated runoff, and helps keep down the number of flies. Several manufacturers have come up with compact manure spreaders that are ideal for small farms. The tractor-driven units carry more manure in one trip and are probably the best way to go if you have a significant number of animals. The ATV units are good for operations with only a few animals stabled regularly or for farms with larger herds that are out on pasture most of the time.

Composters: Compost is an almost magical substance. Break down organic waste, such as food scraps, leaves, grass clippings, and manure, in a compost pile, and you have the best soil amendment in the world—an absolute must-have for a gardener. You can purchase a composter, or you can construct one out of wood or old tires.

Tools

Hand tools: Every farm needs to keep a variety of hand tools. Companies such as Sears (with its line of Craftsman tools) and Snap-On (available at many auto-parts stores) make hand tools that will last for generations and that come with lifetime warranties. At the top of my list of essential items are 25-foot and 100-foot tape measures, a good claw hammer with a comfortable grip, an electrician's pliers, a socket and driver set, adjustable wrenches in several sizes, and a screwdriver set with both flat-head and Phillips-head drivers.

Electric tools: There are dozens of power tools available at hardware and home stores, but there are two must-have electric hand tools for any type of around-the-farm construction project: a seven-inch circular saw and a drill/driver (preferably the type with a rechargeable battery). Need to build your own compost bin? Out come the saw and drill. Hanging a new barn door? Get the saw and the drill. With the range of projects you'll find to use these two tools on, invest in the best and the heaviest duty models you can afford.

If you plan to do a lot of heavy construction and remodeling (thinking about adding a sunroom to the house or building a new barn?), consider adding a heavy-duty reciprocating saw, a chop saw or miter saw, and a jigsaw to your electric tool collection. Other electric tools aren't absolutely necessary unless somebody gets seriously involved in woodworking as an avocation. So, at least initially, save your storage space and money and bypass all of those table saws, routers, and other cool-looking tools on display at the store.

Build your own compost bin to complement, not detract from, your garden or landscaping.

Fencing tools: Farms and fences go hand in hand, and many farmers have gone from the traditional treated wooden fence posts to metal T-posts. To drive T-posts into the ground, you'll need a fence-post driver. Drivers are heavy iron tubes with handles on the sides. You place the open end of the tube over the top of a T-post, put in some earplugs to deal with the noise, hold onto the handles, and pound your post to the desired depth by repeatedly raising the driver and bringing it down quickly on top of the fence post. A pair of fence pliers is a great and fairly inexpensive specialty tool; it's really several tools in one—pliers, wire cutters, a staple puller, and more.

We are also fans of electric fences—they are economical, efficient, and safe, and they come in conventional and solar models. Most stock, even cattle who have a tendency to crowd fences, respect the jolt that these fences can give. If you have an electric fence, you also need a fence tester unless you enjoy testing it the hard way: by touching it yourself. One old farmer tests his fence by holding a piece of grass against the wire. If he feels a "tingle," he knows it's working.

A sturdy garden hose that reaches the areas you want to water is a worthwhile investment.

Miscellaneous tools: We always have our pocketknives handy when we're out working. A "Leatherman" type of all-in-one tool is also useful to carry. You'll find many uses for a round-pointed shovel and a good spade. A pickax/grubbing hoe combination comes in handy, especially if you have rocky soil or stubborn roots, and you'll want a good-quality crowbar as well.

Garden hoses are a must if you want to water anything. Purchase the longest and strongest available (fall is a great time to find really good buys on garden hoses), and resist the temptation to skimp on hoses—there are few things as irritating on a daily basis as a hose that tangles and kinks every time you look at it. The good ones come with long-term guarantees (often twenty-five years or lifetime). We save the guarantee card with the receipt stapled to it because under farm conditions, a "lifetime" can turn out to be only a few years.

Another important miscellaneous tool that you should buy is a digging bar. Five or six feet of hardened steel, a digging bar can be used for levering fence posts out of the ground or vehicles out of the mud as well as a number of other functions. Once you use a digging bar, you'll wonder how you lived without one! We use ours for breaking ice in winter; for digging large rocks out of gardens, construction sites, and fence holes; and for prying apart incorrigible packages.

All About Tractors

For people who are new or aspiring to farm life, tractors are often central to their dreams. They see themselves puttering across a field, baling hay, spreading manure, or planting and caring for a crop. However, if you've never run a tractor before, you'll find that tractors can be a bit intimidating, requiring familiarity with a unique language and skill set that few urbanites possess unless they have farm roots. The following pages will help you learn what you'll need to know before you set out to shop for the perfect tractor.

Horsepower

Since horses provided power for pulling loaded wagons and plows long before people invented engines, early tractor manufacturers rated their products on their ability to do work that was comparable to a horse's work. For example, 1 horsepower (HP) is approximately equivalent to 33,000 foot-pounds per minute. Another way to think of it is that one horse could typically pull

Tractor Talk

Three-point hitch: A standardized mechanism for attaching implements to tractors. The three-point hitch can raise and lower implements. Thanks to a hydraulic pump that delivers oil under pressure to a control valve, it can also hold implements in a desired position.

Power takeoff (PTO): A standardized shaft that spins, providing power to implements from the tractor's engine. In the United States, PTOs are standardized at 540 or 1,000 revolutions per minute (RPM). Older tractors often have a PTO that is driven by the tractor's transmission; new tractors use an independent hydraulic system to operate the PTO. Even with these independent systems, which are significantly safer than the older units, the PTO is dangerous when spinning. Keep the safety shield on the tractor, wear tight-fitting clothing when using the tractor, and make sure that the shaft has completely stopped turning before you get near the unit.

Torque: This is the measure of twist that an engine can apply to the driving wheels or the PTO. Let's say, for example, that you have two tractors advertised with the same horsepower, but one has higher torque. The higher torque model has an engine that runs slower to create the same amount of turning action, thereby using less energy than the lower torque model to do the same amount of work. In practical terms, this means that you won't have to rev the engine as much, the engine is less likely to stall when working under a load, it is quieter, it will use fuel more efficiently, and it will typically last longer.

a 3,300-pound wagon a distance of 10 feet in one minute; therefore, a 10-HP tractor can apply 330,000 foot-pounds of power per minute.

When it comes to modern tractors, there are three different parts of the tractor that manufacturers and dealers talk about when they discuss horsepower. The first is *engine* or *gross horsepower*. It represents the amount of power that the engine produces, but the engine loses some horsepower simply through its own operation. The second part of the tractor that manufacturers rate is *residual horsepower* after engine loss. This is known as the *PTO*—or *power takeoff*—horsepower. The PTO is the device that powers the moving parts on farm implements such as mowers, combines, manure spreaders, and balers, and it is critical to doing work with a tractor. The third way that manufacturers rate horsepower is at the *drawbar*. A tractor's ability to do tillage-type work, such as plowing and disking, is affected by its drawbar horsepower, which steps down from the PTO horsepower. PTO horsepower is usually about 90 percent of gross horsepower, and drawbar horsepower is 75–85 percent of the PTO horsepower.

A tractor's actual performance may be lower in some circumstances than its rated capability. All tractors lose some power at higher elevations; the rule of thumb says that for every 1,000 feet above sea level, there is a 3 percent reduction in horsepower. Tractors also lose power when operating on wet ground and on hilly terrain and when tilling in heavy soils. Over time, the tractor also loses power as wear and tear takes its toll on the engine. This will be even worse if you ignore routine maintenance.

Categories of Tractors

There are three general categories of tractors that are appropriate for small and hobby farms: lawn and garden tractors, subcompact and compact tractors, and utility tractors.

Lawn and garden tractors are light-duty tractors that range from about 10 to 25 HP. They are relatively slow, and most use gasoline, although a few offer diesel engines. They are suitable for mowing grass on an acre or two but are not good for mowing many acres or working over rougher terrain. They can tow light loads, and some are suitable for operating small, ground-driven implements, but few have a three-point hitch or PTO, so they can't operate a wide variety of implements.

Subcompact and compact tractors are built for heavier duty than their lawn and garden counterparts, with engine horsepower ratings ranging from 15 to 50. They were introduced by Kubota in the late 1960s and are now manufactured by several other companies. Intended for the consumer market— designed with comfort, ease of operation, and low maintenance in mind—they also act as a bridge between lawn and garden-type tractors and compact tractors. They accommodate a range of implements and have a two-speed transmission. Compacts (also called estate tractors) are larger than the subcompacts and are very versatile; top-end models may be appropriate for baling hay on a small acreage with a small-square baler.

Lawn tractors are best for mowing and towing small loads.

Most compact tractor models come with diesel engines and offer four-wheel drive, making them good buys when they will have to operate on hills or plow snow. They're faster than their lawn-and-garden counterparts, and they can do more and heavier work. All models come with a three-point hitch and PTO, so they can operate a wide range of implements.

Larger tractors can handle larger projects and implements.

Utility tractors are heavy-duty units, generally designed for commercial-scale farms. They have PTO horsepower in the 45–110 range. As with compact tractors, manufacturers offer most new models with diesel engines, and four-wheel drive is a readily available option. All utility tractors have a three-point hitch and a PTO. For farms up to about 160 acres, these are the workhorse tractors. On larger commercial farms, they do many of the jobs that don't require big power, such as raking hay or hauling round bales to feeders. The smaller units readily operate a small-square hay baler, and the top-end models can easily operate a round baler or "big-square" baler.

The Implements

A tractor by itself is like a foot without a shoe: It can go places, but there's a limit to the kind of hard work it can do. The implements are the shoes—or maybe "magic work boots" would be a better term. Implements take your tractor from being just an engine and wheels to being a do-anything rig. Among the implements for tractors are mowers, tillage tools, loaders, blades, backhoes, fencing tools, and snowblowers.

Alternatives to a New Tractor

You can run a hobby farm without a tractor. A lot of folks do it. This saves the initial financial outlay and gives you time to really hone in on what you need for your particular place. A tractor, or lack thereof, need not be the determining factor of whether or not you can afford the farm of your dreams. Instead, you can hire someone to do the tractor-related work, which does tend to be predominately seasonal. Some farmers will cut your hay on shares. Others are glad to earn a bit of extra income by brush-hogging your field(s) or plowing your driveway as needed. Once we hired a neighbor to plow a cornfield that we were putting into permanent hay and pasture; the neighbor's big rig did in a day what would have taken Ken a few weeks with our equipment.

For occasional projects, you can also rent tractors and implements from rental supply companies. Renting also works well if your own tractor and implements are too small to undertake a one-time big job. We've rented equipment frequently over the years. A used tractor could be a good idea, but just like a used car, a preowned tractor can either be a great buy or a continual nickel-and-dime-you-to-death disaster, so use caution in acquiring one. Often, older, smaller utility tractors in good running condition are available for decent prices. This would be a great time to talk with your farming neighbors and perhaps even see if one is willing to come with you to look over your potential purchase.

One thing to watch for with older tractors is whether the power takeoff is a "live" PTO. If so, it takes a lot more skill to run the various farm implements you'll be connecting to it. (See "Tractor Talk" sidebar.)

We've had good experiences with tractors that are older than we are, but when purchasing used models (even newer used models), expect to incur more repair and maintenance expenses. If you aren't mechanically inclined or don't have the time to do a lot of maintenance and repairs, you might reconsider buying a used tractor. When shopping for older units (and there are many fifty-year-old and older tractors still hard at work), stick to brands that are common in your area of the country. And find out if parts and service are still available from a reputable local dealer. Tractor tires are expensive, so be sure to look at the condition of the tires. Check fluids for condition and quantity.

There are alternatives to purchasing a tractor if money is an issue. On small hobby farms, an ATV or utility vehicle (for example, a John Deere "Gator"), may be a better, less costly choice than a tractor. You'll find a variety of implements available—from manure spreaders to snowplows and hydraulic front-end loader attachments—that work with these versatile machines.

Gray-market (imported brands of) tractors are a special class of used tractor. Gray-market dealers import units for resale into the United States without the consent of their trademark owner. For example, Kubota produces a line of tractors for the United States. Kubota designed the tractors to work correctly with American implements and to meet U.S. environmental and safety standards. Kubota produces a separate line of tractors intended for use in Japan. That line of tractors runs implements found in Japan (which often run at higher PTO speeds than U.S. implements) and that meet different standards required in Japan. Kubota doesn't import the Japanese models into the United States, but gray-market dealers purchase used tractors in Japan, bring them to the United States, and sell the tractors to farmers. These tractors sell for less initially than comparable used American-market tractors, but the original manufacturers don't provide parts, service, or support of any kind. Purchasing a gray-market tractor may seem like a money-saving step up front, but it could cost you far more over time.

Mowers: There are many kinds of mowers available, and the best one for you will depend on your circumstances. A "belly mount" is a mid-mounted mower that sits under the center of the tractor, and there are also rear-mounted and front-mounted units. Finish mowers give a manicured look to lawns. Rotary cutters, also known as brush or bush hogs, are heavy duty—designed to cut thick grass and brush. A sickle mower extends to the side of the tractor behind the back wheel and is used to cut hay and/or clip weeds and grass off at a desired height.

Tillage tools: Farmers use cultivators, tillers, discs, plows, and harrows to prepare seed beds in gardens or fields and to reduce weed pressure during the growing season. For small gardens, a hand tiller is generally sufficient, but for larger operations of a half-acre or more—for example, a sweet corn or pumpkin patch—you'll need either a PTO-operated tiller or a combination of a plow, disk, harrow, and cultivator. The plow does the first rough turning of the soil, the disk smoothes it over, and then the harrow does the fine smoothing. Farmers use cultivators in between rows while crops are growing in order to reduce weed pressure.

Front-loading attachments can help with moving bales of hay, among many other uses.

Loaders: A loader is the tool of tools. It allows you to dig and scrape; to move materials such as hay, manure, compost, or snow; to grapple equipment or machinery by lifting heavy items with a chain; and to do light grading of roads and driveways. The loader fits on the front of the tractor and comes with a bucket. Other attachments are readily available, including a fork for spearing and moving big round bales of hay and a forklift for raising pallets and other flat items. Modern loaders not only lift but also have powered control for downward movement and tilting capability.

Blades: Blades are available as rear- or front-mounted implements. They are good for heavier scraping jobs, including grading roads and shaping ditches, and they can drag soil or gravel for short distances. Some blades are straight, but angled blades, although a little more expensive, are far more versatile and worth the up-front cost.

Backhoes: Backhoes are rear-mounted digging accessories that make digging a snap. Depending on the size, you can excavate fairly large holes in rough ground in a short time; for example, digging for a building's foundation or a septic system is a reasonably easy chore with a backhoe. That said, unless your operation calls for a lot of this kind of work or you have a bundle of cash that you can't wait to spend, a backhoe isn't necessary for farm life. You might go several years without really needing one, so unless you can't live without one of your own, consider renting one or paying someone else to do the job when the occasion arises.

Fencing tools: Post drivers and posthole diggers/augers are available for tractors. They make large fencing projects go much faster than if you were to do it all by hand. Posthole diggers/augers also come in handy for constructing pole buildings and for planting trees and shrubs.

Snowblowers: In most areas of the country, a loader bucket and/or blade will take care of snow, but if you happen to live in a place known for heavy snowfalls, a blower will cut through the drifts with the greatest of ease. It will also take much less time than it would take to move the snow with buckets and blades.

Other PTO-driven tools: Many other tools are available for your tractor, and your needs depend on your workload. There are seeders and planters; manure spreaders; feed augers, grinders, and mixers; cement mixers; and wood splitters and saws. Name a task, and there's probably a tool to make that chore easier.

4

Nature's Troublemakers and Farm Safety

Yesterday afternoon, Ken and I watched about one hundred antelope having a grand time playing on the hill behind our house. Then, last night, we stood on the deck under the stars, listening to an owl hoot in the trees on the ridge next to the house. He asked us, "Who…? Who…? Who…?" We couldn't answer him.

Counting wildlife encounters as part of our daily lives is what makes this life so appealing to us, but there are some encounters we're less fond of. Weeds, predators, and a host of other troublemakers think of our place as a fine home or a good place to visit; they likely will think of your home much the same way.

Weeds

To quote Emerson, "A weed is a plant whose virtues have not been discovered." Undoubtedly, when it comes to most noxious weeds, such as thistle, knapweed, and leafy spurge, it's hard to imagine them having any virtues at all. But they do, as butterflies would tell us if they could talk. Nevertheless, when weeds are on your land, it's hard to share the butterfly's view of the world. So what can you do?

Completely eliminating weeds is an impossible task, even with herbicides (as evidenced by the fact that we have more acres of weed-infested land today than we had before the introduction of chemical herbicides). But control of weeds is a possible and laudable goal. The type of crops you

A rototiller can be a helpful tool in the garden for those who want to stick to natural means of weed control.

want to grow and the kinds of weeds you are trying to control will help determine which approach(es) you choose for weed control.

When you opt to work on a control strategy for weeds, it needs to be a long-term commitment; it can take three to seven years of consistent application of controls (mechanical or biological) to have a significant impact. But the good news is that if you make the commitment, you can succeed. There are many different weed-control methods, and great informational resources abound, so here is an overview to launch the subject.

Mechanical Techniques

Many weeds are susceptible to mechanical techniques, including pulling, mowing, hoeing, cultivating, and burning. Pulling weeds is still the most effective method of control for small areas or limited infestations and is done most easily when the soil is slightly moist. During our first summer on the Kremmling farm, the loco weed (which can be toxic to livestock and wildlife when eaten in sufficient quantities) had a bumper crop throughout the mountains of Colorado, thanks to a moist spring. Most people spray to get rid of it, but the moist spring made for easy pulling, so I opted to spend an hour or so each afternoon for a couple of weeks pulling out the biggest patches, which kept the weed down to a minimum. Then, because we hadn't sprayed, shiny green beetles showed up in August. Those beetles ate all of the remaining plants. The next summer, the beetles arrived again and cleaned up any loco that tried to sprout on our place. Those who had sprayed before had to keep spraying because they had not eliminated the roots (which pulling did), and they didn't have the helpful beetles cleaning up for them.

For pastures, fields, and corrals, check with your local extension office for up-to-date information on how to best eradicate/control unwanted species. In most areas, the Department of Conservation may be willing to come out and look at your ground, assess the vegetation, and make recommendations for management. For gardens, a good, sharp hoe makes quick work of weeds, and if you have a rototiller, shallow tilling between rows can shorten the task. The key to hoeing is to do it frequently; don't let weeds get too thick or too large before you nip them with the hoe just below the soil surface, or it will take a whole lot more muscle power to eradicate them.

Be aware that weed seeds can lie dormant for years, waiting for the right conditions to call them to life. Hoeing or tilling tends to stir up that seedbed, bringing ungerminated seeds to the surface, where they can sprout. This is not bad, as long as you don't neglect that new crop of weeds, realizing that weed control is progressive—the more weeds you can eradicate before they go to seed, the fewer seeds enter the seedbed. The more often you can hoe/till/pull the new little weeds that sprout, the fewer that will come to maturity to add more seeds to the ground. Eventually, weed seeds will be more or less exhausted from the seedbed.

Stubborn perennial weeds that spread by roots as well as seeds will need to be dug out or pulled out by those opportunistic roots. Hoeing them off at ground level will not eradicate these rascals. One example is quack grass, which, in addition to seeding, sends out rhizomes from its spreading roots. Each rhizome can sprout into a new mound of grass, so the potential exists for a baby bunch to be born every few inches.

Mulch can help control weeds, enrich the soil, and improve your garden's aesthetics.

Controlling Weeds

For more information on controlling weeds, check out the University of Georgia Center for Invasive Species and Ecosystem Health website at www.invasive.org. Clicking on "Plants" will take you to various categories of invasive plant species that you can explore. Clicking on each individual species will take you to photos and more information about that species.

Types of Weeds

- Summer annual weeds grow every spring or summer from seed. These weeds produce seeds, mature, and die in one growing season. Their seeds germinate mainly during a two-month period in late spring and early summer, and the seeds are dormant until the next spring. Thanks to their self-seeding nature, they may seem like perennials.

- Winter annual weeds grow in late summer or fall from seed, and they then mature and produce seeds the following summer. Some, including chickweed, can germinate under snow.

- Biennial weeds grow from seed at any time during a growing season. They normally produce a rosette of leaves close to the soil surface in the first year, and then they flower, mature, and die during the following growing season.

- Perennial weeds establish themselves in new areas by seed. Once established, they can proliferate from the roots. They live for more than two years. Because they can grow from pieces of root, perennial weeds may take off after soil preparation and cultivation because their roots have been chopped and distributed. Most perennial weeds that spread from extensions of their roots grow in circular patches if left undisturbed. In cropped fields, the patches are often oblong, following the pattern of cultivation. Well-established perennials are the most difficult weeds to control.

The ubiquitous dandelion is an annoyance to farmers across the country.

Rototilling quack grass actually spreads the weed because the tiller blades chop the roots into little bits and pieces, each of which are fully capable of sprouting wherever they land. Mulching will not prevent the rhizomes from sending out their new growth under the mulch. It'll be a bad day if quack grass makes its insidious way into your asparagus bed or row of raspberries. Be militant about manually preempting the spread of this weed, or you may have to use chemicals to gain the upper hand.

Another aid to weed control is heavy mulching. In addition to conserving water in the soil, thus reducing the need to water quite as much, it keeps weed seeds in the dark, preventing them from germinating. A plethora of gardening sites, books, and magazines expound the benefits and types of mulch, and you should be able to settle on one or more that work well for your particular budget and region. *Cover crops* (such as buckwheat, for example), also called "green manure," planted in dormant areas can also act as mulch. When tilled into the soil at the proper stage of maturity, these crops not only enrich the soil but also help prevent weeds—those perpetual opportunists—from taking over open ground by shading and crowding them out.

Mowing and Controlled Burning

For areas such as pastures, fence lines, roadways, orchards, and any other zone where the land is in a permanent cover crop, mowing and controlled burning are both viable approaches in the war against weeds. Mowing, weed-eating, and brushhogging work best when the weeds are just ready to bloom but before more than the first few flowers are opening. This timing is absolutely critical. If you cut before the flower heads form, the plant has plenty of spare energy left in its roots; it

Grazing goats can act as natural lawn mowers when it comes to fighting weeds.

simply sends up a bunch of new shoots and a new crop of blooms. If you cut after the flowers start opening in earnest, the seeds will continue to mature and spread, even though the weed is lying dead on the ground (why are weeds so tenacious?). But, if you mow just as the flowers first open, the plant has put most of its reserved energy into flower production and will have a tough time regrowing in time to create a secondary bloom.

You can use grazing animals for "mowing," but, for effective weed control, you can't just turn a herd loose and expect that they will go after the weeds efficiently; managed grazing is necessary. Sheep and goats are often the top choice for this task, but other critters will afford some help as well, depending on the species of weeds. For example, ordinarily, cattle won't touch thistles, but Gregg Simonds, a clever rancher in Utah, decided to try spraying Canadian thistle patches with molasses and water before letting his cattle into an infested paddock. The result: the cattle "ate the thistles right into the ground."

Horses don't tend to be much help when it comes to weed control. They may eat a portion, but they like to go for tender leaves and shoots. Typically, rather than treating all vegetation with equal interest, they return to areas they've previously grazed down, looking for new growth and leaving hummocks of weeds and grass to toughen and send out seed.

Controlled burning is usually done in the spring. It is a field-scale fire kept burning at low temperatures, intended to take all vegetation back to the growth point and give grasses and legumes a chance to compete with the weeds. If a weed infestation is too serious, you may require a hotter burn, followed by reseeding of grasses and legumes.

Timing is important. Burning when unwanted species have begun to grow but before wanted species have taken off helps give the latter a better chance to outgrow, and thus outproduce, the weeds.

Do not embark upon burns lightly; even a light wind can turn a managed burn into an uncontrolled conflagration. Be sure to check with your fire department about burning because you may need to obtain a permit. You can also avail yourself of your local Department of Conservation to help with a controlled burn. They usually sponsor regular classes on how to conduct a successful burn, and, in some cases, they will physically assist you in carrying out a burn on your own land.

In early spring, when weeds are just emerging, the best mechanical approaches in field crops are cultivating and/or spot burning with weed flamers. You can use either technique several times before the crop is well established. You'll find a number of cultivator styles designed to dislodge, cut, or bury plants, or all three. Be aware that any mature seeds that you till under will add to your next crop of weeds, so be sure to bury them before they flower, and, again, if the weeds are of a type that spreads by rhizomes, avoid tilling and cultivating.

Biological Techniques and Herbicides

Biological weed-control techniques use beneficial insects ("beneficials") or plant-specific disease organisms for controlling weeds, but each insect or disease tends to be specific as to which weed(s) it will control. Using beneficials for weed control is less expensive than using herbicidal sprays, and, once a population of organisms is established, minimal effort is required to conserve it. Using beneficials is also less disruptive

Certain types of "beneficials," such as the green lacewing, benefit the garden by eating insects that can damage plants.

ecologically, enabling you to maintain natural biodiversity. Many states are now running beneficial insectaries, where you can purchase insects for some of the most troublesome weeds in your area. The key to using beneficials is to provide them with good living conditions, which means avoiding the use of chemicals that kill the good bugs as well as bad.

Herbicides should be the last choice in weed control or at least limited to new infestations of fast-spreading nonnative weeds. Native plants adapted over millions of years to fill unique ecological niches, whereas nonnative weeds developed in other countries or regions. Insects, disease, and competition with other species keep weeds in check in their native environments, just as they do our native plants. Without their natural enemies, though, many of these nonnative weeds become highly invasive when they get to the United States. Spreading like wildfire, they reduce the diversity and quantity of native plants. Federal and state agencies list fast-spreading nonnatives as *noxious weeds*, those that are "competitive, persistent, and pernicious." Learn about noxious weeds in your area and take steps to control them.

For best results, combine herbicides with other techniques. For example, cutting followed by herbicide applications, or controlled fires followed by spot applications of herbicides, work better than just spraying. The problem with relying on herbicides is that they will reduce natural biodiversity, which results in more problems over time. And weeds, ever adaptable, can develop resistance to herbicides, thus calling for different and sometimes more disruptive chemicals.

Coyotes are becoming increasingly problematic for rural and suburban farmers alike.

Raccoons are notorious food thieves and will also prey on your fowl if they gain access to the chicken coop.

Predators

The US Department of Agriculture (USDA) estimates that predators (wild and domestic) account for almost $100 million in losses to farmers and ranchers annually. So, if you have livestock and pets, you may have good reason to fear for your animals' safety. However, according to Dr. John Shivik, a wildlife biologist at the USDA's Predation Ecology Field Station in Utah, there are ways to reduce the impact of predators on your livestock. Do keep in mind, however, that predators are important components of ecosystems. In nature, predators feed not only on large herbivores but also on lots of small rodents and rabbits (in fact, several studies show that rodents make up about 90 percent of the diet of coyotes). They'll also eat insects and carrion. If you consider the population explosion that would take place in the mouse or insect world were we to remove coyotes from the ecosystem—to say nothing of how many rotting carcasses would dot the countryside—you'll see that moderation is easily the wisest choice.

When they do kill livestock or pets, predators aren't trying to ruin your day, cut into your profit, or break your heart; they're simply doing what they were born to do. No one is putting out a food bowl for them morning and night, they have no concept of property lines, and they see no difference between wild and domestic prey. As Dr. Shivik says, "Predators kill things for a living; that is their job." They don't kill because they like it, but because they must eat to survive.

It's the farmers and ranchers who live in remote areas who have the majority of problems with wild predators. In small farm towns and rurban areas, domestic dogs tend to do most of the damage, although coyotes are moving into the 'burbs, finding it an easy place to make a living (they've even been seen dodging traffic in the Bronx).

Wolves and even domestic dogs can be dangerous predators.

Fido and Spot don't have to be wild, vicious, or even brave to chase sheep or kill chickens: They're simply following their natural impulses. They are capable of carrying out the hunting sequence of their forebear, the wolf—from orienting and tracking to stalking, chasing, herding, attacking, and killing—although they usually do so for sport rather than survival.

Not all predators kill livestock. When they do, they tend to be opportunistic killers, seeking whatever is easiest to meet their needs. Why expend a lot of energy hunting wild and wary prey when a whole coop full of captive fowl is close at hand? Raccoons and possums, too, are finding that raiding trash cans and cat- or dog-food bowls in the suburbs fills them just as full as chasing prey in the wild.

As a rule, predators go for young, old, weak, or sick animals first. As they become desperately hungry, though (such as during a drought), they become much more aggressive and will attempt to take healthy, mature animals. Since healthy animals suffer less predation, good feed and adequate health care can help ensure that your animals are not fair game.

There is no magic answer to all predator situations; each predation event includes unique circumstances and thus will require unique responses. Overall, the best approach to protecting your livestock (and pets) is to make predators think that eating at your house will be harder than chasing mice, rabbits, or deer. You can do this by developing an understanding of predators and learning to apply nonlethal techniques that reduce predation.

Although Wile E. Coyote may have looked the fool in his encounters with the Road Runner, he's not a good example of the species or of predators in general. Since it is the "job" of predators to kill, they are intelligent, curious, and, most of all, adaptable. Consequently, changing the behavior of predators—although it may be possible—is harder than changing your own.

Becky Weed is a sheep producer who has learned how to successfully adjust her behavior. Becky and her husband, David Tyler, didn't grow up on farms, but, in the 1980s, they started raising sheep in Montana and have been farming full time since 1993. Early on, Becky and David lost 20 percent of their flock to coyotes. They called an Animal Damage Control agent, who shot and trapped a few coyotes, but, as Becky says, "We knew we couldn't kill all the coyotes that came through, and we didn't want to, even if we could." She has since built a very successful marketing strategy on the concept of offering "predator-friendly" wool. As founding members of Predator Friendly®, a co-op that brands and markets environmentally conscious products, Becky and David have seen their business grow steadily. Predator Friendly®'s website (www.predatorfriendly.org) lists the following practices by which their certified producers abide:

- Using guardian animals such as llamas, donkeys, and dogs to keep livestock safe
- Scheduling pasture use when predation pressure is low
- Grazing cattle with smaller livestock to protect sheep, goats, and calves
- Timing calving and lambing to avoid predation risk
- Lambing in sheds, secure fenced lots, or protected pastures
- Making frequent and unpredictable patrols in pastures
- Protecting vulnerable animals by fencing out predators
- Learning the ecology and habits of area wildlife

Identifying Predation and Predators

The first step is to identify the predator, but keep in mind that sometimes predators get a bum rap. A farmer comes upon the corpse of a dead animal, and, because there are obvious bite marks, he or she assumes that a predator killed it. But animals die from a number of causes, and unless you see the predator in the act of attacking a live animal, you don't know if a predator killed the animal or if the animal died from natural causes, with predators simply scavenging afterward. It's nature's cleaning system at work.

When you suspect predator damage, assess the scene. Signs of a struggle, like drag marks, torn hair, wool or feathers left on brush or fences, or blood spread around a large area all point to predation. If there are no signs of a struggle, it may help to examine the carcass. When a predator feeds on a dead animal, that animal will not bleed under the skin at the bite marks. This type of bleeding, known as subcutaneous hemorrhage, is only present when the animal's heart was beating at the time of the predator attack.

When signs of struggle or subcutaneous hemorrhage are present, the next step is to try to confirm the kind of predator. Each species leaves its own telltale signs at a kill. For example, canid species (coyotes, dogs, wolves, foxes) tend to attack from the sides and the hindquarters, grabbing prey under the neck, whereas cats tend to jump up on the back, biting the top of the head or back of the neck. Close examination of paw-print size and shape, tooth spacing and size, feeding habits, and pattern of killing help correctly identify the predator responsible for the kill.

Guardian Animals

For thousands of years, farmers in Europe and Asia used guardian dogs to protect their sheep and goats, as did Americans. However, over time, farmers switched from using guardian animals to

The Great Pyrenees has a long history of guarding flocks, being bred and raised for the task.

protect their flocks and herds to using guns, poison, and traps. Today, farmers like Becky and David are showing that the old approach of using guardian animals is still practical.

Killings usually occur at night or in the very early morning, when people are normally asleep. A guardian animal is on duty twenty-four hours a day, are alert and protective during these hours of greatest danger. Dogs are probably the most common guardian animals, but farmers also use donkeys, ponies, mules, and llamas to protect sheep and goats.

You'll need to raise guardian animals very differently than you would raise pets. While they are still puppies, guardian dogs need to bond to the animals they are going to protect, not to the family members. At the same time, you must be sure to handle them often enough so that you can safely feed them, take them to the vet, chain them when you will be working with the flock, and perform other necessary tasks with them. Certain breeds, such as the Great Pyrenees, Anatolian Shepherd, Maremma Sheepdog, and Akbash Dog, are characteristically used as guardians. Research the various livestock guardian breeds to help you decide what type of dog would work best for your particular operation. The USDA also has publications on livestock guarding dogs at the following link: www. nal.usda.gov/awic/companimals/guarddogs/guarddogs.htm.

When coyotes and domestic dogs are the problem, one or two guardian dogs are sufficient to protect a farm flock, but if wolves or other large predators are the major concern, dogs may or may not be effective. Although some farmers and ranchers report success warding off large predators with three to five dogs, Dr. Shivik notes, "In the western United States, guardian dogs are often killed by large predators, particularly wolves."

Donkeys and llamas, which live longer than dogs and don't require special feed, really dislike coyotes and dogs but tend to be scared of, or even vulnerable to, larger predators such as bears,

OUTSMARTING COMMON PREDATORS

Predator	Where	Domestic Prey	Characteristic Behavior
Coyote	Traditionally found in rural areas west of the Mississippi, but they are extending their range to towns, cities, and areas east of the Mississippi, with a significant increase in the Southeast	Sheep, goats, calves, poultry, and small domestic dogs or cats	Hunts alone, in pairs, or occasionally as a family pack
Wolf	Found in pockets of the West and around the Great Lakes	Capable of taking mature cattle, llamas, and horses, as well as all small stock	Hunt in a pack
Fox	Throughout the country; often live in towns	Mainly lambs/kids or poultry	Hunts alone or in pairs
Domestic Dog	Found anywhere inhabited by people	Capable of taking mature livestock as well as all small stock	Hunts alone or in packs
Bear	Found in remote areas and wildland/urban interfaces over much of the country	Capable of taking all classes of livestock	Hunts alone or with cubs
Bobcat	Found mainly in remote areas over a fairly large portion of the country, but they are seen in largest numbers in the western states	Small stock, poultry, domestic dogs and cats	Hunts alone
Cougar/Panther	Found mainly in mountainous regions of the West; the South has remnant populations of native panthers	Capable of taking all classes of livestock	Hunts alone
Birds of Prey	Found throughout the country	Small stock, poultry, domestic dogs and cats	Hunts alone

Attack Pattern	Feeding	Fencing
Attacks from sides or hindquarters. Bite marks and subcutaneous bruising are often found under the neck and throat of prey, as well as bloody foam in the trachea. Usually attacks right before dawn or right after dusk.	Usually begins on the flank, just behind the ribs, consuming organs and entrails	6 feet high, with additional 3 feet buried underground, or 5-foot-high electric fence, with wires starting low and fairly close together
Similar to coyotes, but large tooth patterns and often multiple kills in one night	Similar to coyotes	6-foot-high woven wire with electric wires along top and bottom
Similar to coyotes but small tooth patterns	Similar to coyotes	4-foot-high net wire with openings less than 3 inches square, buried to height of 3 feet with a 1-foot apron
Indiscriminate mutilation of prey, bite marks are found on multiple areas of prey's bod. Often attacks during the day	Often kills large numbers of animals at one time but does very little feeding	Same as coyote fencing
Kills with crushing bites to the spine, skull, and dorsal side of neck of prey. Claw marks are often found on the neck, back, and shoulders of larger prey. Often kills more than one animal.	Consumes the udder and flank and removes the paunch and intestines intact; carcass may be almost entirely consumed. Often drags its prey to cover and sometimes covers it with grass and dirt.	Electric fence at least 3 feet high
Usually kills small animals by biting on the head or back of neck. Often leaps on the back and bites the neck and throat of larger prey. Hemorrhaging from claw punctures often can be found below the prey's skin on the neck, back, sides, and shoulders. Paired upper and lower canines are usually spaced ¾ inch to 1 inch apart.	Often begins feeding on the viscera after entering behind the ribs. May drag and cover killed prey.	5-foot-high woven wire
Usually bites the back of the neck and skull, causing massive hemorrhaging. Large canine tooth punctures are found; upper canines are spaced 1 ¾ to 2 inches apart, and lower canines are spaced 1 to 1 ¾ inches apart. Large claw marks are found on the head, neck, shoulder, and flank of prey.	Usually eviscerates the carcass, removing the entrails and moving them aside. Consumes the lungs, heart, liver, and larger leg muscles first. May drag and cover the carcass.	9-foot-high heavy woven wire or electric fence
Often kill poultry or small mammals (new lambs and kids are fairly vulnerable). Talon punctures are found on the prey's head and body, along with internal hemorrhaging caused by talons. Tufts of scattered feathers, wool, or hair are found; the carcass is often "skinned out." Streaks of white feces are seen in the immediate area of the carcass.	Consumes the entrails and organs; sometimes opens skull and eats the brains. Removes the ribs near the spine on young animals.	Confine outdoor animals in wire-topped cages. String scare balloons or aluminum pie pans on poles.

Fences designed to keep animals in may not always keep predators out.

mountain lions, or wolf packs. In the United States, you can purchase donkeys and llamas for less than what you would spend on a good livestock guardian dog.

As with dogs, it is best to purchase donkeys or llamas when they are young and raise them with the flock or herd that they will protect. A single female or single gelded male is less likely to harass the animals it is meant to protect than are multiple guardian animals or ungelded males, and a single donkey or llama will stay with the flock for companionship. However, if you use a herding dog for working your flocks and herds, the donkey or llama may perceive the dog as a threat and interfere with its ability to work.

Whichever type of guardian you're considering, remember the following:

- The guardian animal needs to bond with the animals that it's protecting, and bonding can take time.
- Introduce the guardian to the animals slowly, across a fence. It's usually easiest to make the introduction in a small area rather than in a large pasture.
- Each animal is an individual and will react differently in different situations. Some individuals don't make good guardians even though the species or breed itself does.

Physical Barriers

Fencing and enclosures are designed to place a physical barrier between predators and their prey. As Dr. Shivik says, "Exclusionary devices can be as simple as an easily strung electrically-energized temporary corral or as complex and expensive as a dingo-proof fence stretching from one side of Australia to the other."

People build most fences to keep livestock in. Fences designed to keep predators out are more expensive, so they are rarely cost-effective for large areas. Night penning is a cost-effective fencing approach that works well, especially for small- and medium-size operations. It involves bringing animals back into a small, predator-proofed area in the evening. Adding lights to night pens increases the pens' effectiveness.

Since young animals are most vulnerable, having babies near the farmstead house reduces predation, particularly if you can move mother animals into a shed or barn as they approach parturition. Buildings are also crucial for keeping smaller animals like poultry and rabbits, which are vulnerable to a wide variety of predators, safe. Timing when your baby animals are born can also influence predation. Lambs, calves, and other babies that are born in late spring and early summer are less subject to predation than those born in winter or early spring because small prey, like rabbits or voles, are abundant in warmer weather, and, consequently, predators are not starving.

Farm Safety

It is a dangerous world, but farms are more dangerous than most other places, so farm safety is one of the most important things to think about. Not to scare anyone, but as Benjamin Franklin said, "An ounce of prevention is worth a pound of cure." If you're aware of the dangers, chances are much greater that you can avoid them. Consider some statistics:

- In 2002, the agriculture industry had the second highest rate of fatalities in the United States due to accidents.
- In an average year, tractor rollovers crush more than 100 American farm workers to death.
- Every day, about 500 US agricultural workers suffer lost work time injuries, and about 5 percent of these result in permanent impairment.
- Farming is particularly dangerous for children. On US farms and ranches, an estimated 104 children younger than 20 years of age die annually of agricultural injuries. More than 30,000 serious injuries occur annually to children under the age of 20 who live on, work on, or visit a farm.

Heavy equipment, such as tractors and combines, can be involved in rollovers, runovers, and in the pinning of people against stationary items. High-speed spinning and chopping devices, such as PTO-driven equipment, can easily catch hair, clothing, and hands. This equipment is also very noisy and can reduce hearing capacity over time if operated without some type of hearing protection.

It can be particularly dangerous to work in enclosed spaces, such as silos and manure pits, which can lack sufficient oxygen or contain toxic gases, some of which are odorless, thus rendering them undetectable until it may be too late. Inhaling even a small amount of such gas as nitrogen oxide can permanently damage the lungs. The National Agriculture Safety Database (NASD) contains information on the very important matter of toxic gases, including safety recommendations and cautions.

Test enclosed spaces with gas and oxygen testers before you enter. If the air isn't suitable, set up an air pump to provide a suitable atmosphere. When working in an enclosed space, wear a safety harness and have another person nearby, outside the enclosed space. The harness allows rescuers to get you out without having to enter the space themselves. It is recommended that you post danger signs at the base and door of your silo to warn unsuspecting or unknowledgeable individuals.

Farmers often handle and store hazardous chemicals, such as herbicides, pesticides, and fertilizers, many of which are known carcinogens. You must exercise the utmost care if you to opt to use such chemicals. Read the directions carefully, wear personal protective

Growing up on a farm can be a rich and educational experience for children when taught and supervised properly.

Children must learn to treat farm animals with care and respect.

equipment (PPEs) as specified by the manufacturer, and dispose of chemicals and the containers they come in according to the manufacturers' instructions.

Livestock hazards are abundant, and the animals don't necessarily have to be aggressive to hurt you. Any animal may kick, bite, buck, butt, or stomp out of fear, stress, or just plain obliviousness. One hobby farmer broke a bone in his forearm during an incident in which a yearling steer was so intent on the feed bucket in its owner's hand that it cornered the farmer, who, lacking a stick or other implement to drive the feed-focused bovine back, fended it off with his arm. He discovered to his chagrin that the skull of a steer is considerably tougher than a bone in a human arm. And to add insult to injury, the rascal didn't have a clue that he'd helped send his owner to the emergency room. He was too busy chowing down on grain.

Big animals have big feet and weigh a lot. Add to that the fact that most don't have a clue that they can crush your foot or smash you against a gate, and it becomes imperative to watch where your feet are at all times. Horses can be taught to respect your personal space, but it's always wise to be watchful around animals.

Breeding males of any species are especially dangerous during the breeding season. In nature, fighting is one of the steps to winning a female, so a normally mellow bull, stallion, buck, ram, or boar can suddenly become an aggressive terror. As a general safety rule, always carry a stout stick or livestock whip with you when you enter the pasture with breeding stock or approach mothers with babies at their sides. Some breeds of cow can be especially protective of their young, so approach only when necessary and always exercise extreme caution.

One Missouri farmer tried to help a weak newborn calf into shelter, and, in spite of a stick-wielding friend trying to fend off the mother cow, the cow managed to knock the farmer down and trample him multiple times before he was able to roll out under the fence. He was able to walk away, but he suffered several broken ribs and major bruises. Others have experienced permanent damage from similar encounters.

Purposefully make yourself aware of the hazards around your farm and make sure that your family, visitors, and workers are aware of them as well. Often those who have not lived in rural areas view a visit to the farm as an adventure of exploration but don't realize the possible hazards. Little children, especially, often have no clue that geese may chase them, goats could butt them, horses can kick them, and cattle or pigs can run them over. And if your own kids are excited to show their city cousins the wonders of farm life and farm animals, they may temporarily forget the safety precautions you've drilled into them, so be sure and give the whole younger contingent a safety speech when visitors come over.

In addition, common sense goes a long way toward removing certain hazards around the farm. Your local fire department or extension agent may be happy to come out and walk through your property with you for a voluntary safety audit. Do not tolerate risky behavior from anyone on your farm. Wear appropriate PPEs whenever working with machines and/or animals. Basic PPEs include hearing protection, goggles, steel-toed work boots, and good leather gloves, but many jobs require specialized PPEs as well.

No matter how much experience you have with animals, always keep safety in mind.

Barn Fires

Although the big disasters garner the majority of the news coverage, the most common disasters for farmers are barn fires. There are a number of steps that you can take to reduce the chances of a barn fire and to improve the likelihood that your animals will survive in the event that fire breaks out.

Barns are usually full of highly flammable materials, such as hay chaff. A stray ash, an electrical short, or a hot engine can set off a conflagration in no time at all. Absolutely forbid smoking in and around the barn, and enforce this rule with guests as well as family members and employees. Be extremely conscious when using electrical appliances or heavy equipment around the barn, and don't leave them on unattended.

Inspect electrical wiring annually. If you use brooder lights to raise poultry, ensure that they are properly positioned and are not in contact with flammable material, such as bedding. And, probably most importantly, don't store damp hay in a barn or haymow. Farmers do their best not to bale hay

when its moisture content is too high, knowing that it is subject to mold and possible spontaneous combustion. But sometimes, in spite of their best efforts, such fires still occur.

Keep fire extinguishers near every entry point and make sure they are up-to-date and large enough to be effective. If you regularly keep animals stalled in the barn, consider the possibility of installing sprinklers. You might also consider investing in a good heat-and-smoke sensor in the barn with alarms mounted outside the barn and inside the house. The sensor will pick up any sudden increase in building temperature, even if there isn't active smoke yet.

If a fire has started, use extreme caution when entering the barn. If you think you can enter safely, evacuate the easiest-to-reach animals first. This is a case in which some predesigned facilities will help: Construct a fenced area far enough away to safely keep the animals but close enough so that you can quickly contain the animals if you are evacuating them from a barn. Consider installing multiple exits so that if flames block one exit, other options are available.

Disaster Preparedness

For most Americans, disaster is something we see on the evening news. Yet, each year, hundreds of thousands of us personally suffer through some type of disaster—floods, hurricanes, power outages, and wildfires, to name a few. Disasters can be even more of a problem in rural areas because rural emergency services are often operated by volunteers and may have to cover large areas.

A spokesperson for the Federal Emergency Management Agency (FEMA) says, "You don't want to be thinking about what you're going to do in an emergency situation for the first time as that crisis is occurring." She adds, "There are three key things we urge people to do:

- Have a plan. We think it's the most important thing people can do.
- Have a disaster supply kit.
- During a disaster, listen to your local emergency managers. These officials will guide you safely through a crisis."

If you have pets or livestock, make sure that your evacuation plan addresses their needs and that you know where you can take them to be safe. Animals are often not allowed in emergency shelters.

If you must evacuate due to an emergency and cannot trailer your livestock, use spray paint or grease marker to write your phone number on their shoulders, sides, or wherever so that whoever finds them can contact you. Then let them loose so they can find their way to safety.

To create a disaster plan, take the following steps:

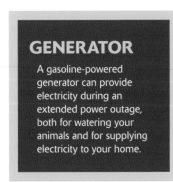

GENERATOR

A gasoline-powered generator can provide electricity during an extended power outage, both for watering your animals and for supplying electricity to your home.

Learn and evaluate. Although no one wants to live in a state of perpetual fear, it is a good idea to familiarize yourself with what types of disasters could occur in your part of the country. Consider both natural and manmade events, including floods, hurricanes, thunderstorms, tornadoes, extreme winter weather, earthquakes, mudslides, wildfires, hazardous-materials incidents, and terrorist attacks. They may never happen, but if you're prepared, you'll be miles ahead if they do.

Your local fire department or county emergency services personnel can help you consider all of the possibilities you might overlook. For example, the fire department may know

Be prepared for the unexpected.

about hazardous chemicals being transported through your community. Some communities have developed their own disaster preparedness and relief chapters. Usually run by volunteers, these organizations are geared to step in give needed assistance during personal or community disasters.

Determine means of contact. Determine how you will make contact with other members of your household if you are separated by an emergency. Identify two meeting places: one near your home (perhaps a nearby tree or telephone pole) and another farther away, in case you cannot return home. In addition, identify an out-of-town contact to act as a family communicator.

Draw up a floor plan. Draw a floor plan of your home. Mark two escape routes from each room, and review it with all members of your household. Practicing fire drills is also a good idea. Children will especially benefit from having practiced what they will need to do in an emergency. Likewise, if you keep animals in a barn, plan escape routes from the barn.

Post emergency numbers. Post emergency telephone numbers by telephones or, if you no longer have a landline, post them near a computer and in a central location like the kitchen. Emergency numbers should include numbers for veterinarians and livestock handlers.

Teach children how and when to call 911. If you live off main highways, post written directions to your home next to the list of emergency numbers so that you can clearly explain to emergency dispatchers how to get to your home. GPS systems are great, but not infallible, and some rural addresses can be hard to find. "Turn left after the first cattle guard, drive three miles, and then turn right after the green mailbox" is not something you want to have to think about while the barn is burning.

Know the location of shutoff switches. Make sure that everyone in your household knows how and when to shut off water, gas, and electricity at the main breakers or valves. If you have to evacuate, and you have time without endangering yourself or your family, fill the bathtub, sinks, and other containers with water before you turn off the water. Then, when you return, you will have water immediately available for cleaning and other uses, even if the electricity is off.

Reduce economic impact. Reduce the impact of disaster on your property and your household's health and financial well-being. For example, review property insurance policies before disaster strikes—make sure that your policies are current and that they meet your needs in terms of type and amount of coverage and hazards covered. Floods aren't covered under normal homeowners' policies unless you have a flood policy. Most "acts of God" are not covered and must be added separately to regular policies.

Reviewing life and health insurance policies is also a good idea, and consider starting an emergency savings account to use in a crisis. It is advisable to keep a small amount of cash or traveler's checks at home in a safe place where you can quickly gain access to it in case of an evacuation.

Inform emergency personnel. Does someone in your household have special needs (someone with impaired mobility, a child who is home alone after school, a person without a vehicle, or another special circumstance)? If so, let local fire and emergency response personnel know what the situation is at your home. Stickers for doors and windows that advise rescue personnel about the presence of indoor pets are available as well.

Consider others. Do you have neighbors who will likely need help? Can you volunteer your assistance in times of trouble? Can you participate in developing your neighborhood's or community's capacity to handle an emergency?

Consider your animals. Where will you evacuate them to if you need to leave? If you face an extended power outage, how much water will they need and how will you supply it?

Preparing a Disaster Kit

- **Water.** Stocking water reserves should be a top priority. It is very hard to ration water in emergency situations. Therefore, it's critical to store adequate amounts of water for your household. Because you will need water for drinking, for sanitary purposes, and for cooking, store at least one gallon of water per person per day in thoroughly washed plastic, fiberglass, or enamel-lined metal containers. Try to avoid milk jugs because milk-protein residue may render the water undrinkable. Sound plastic containers, such as soft drink bottles, work well. You can also purchase food-grade plastic buckets or drums.

   FEMA recommends that if you do not purchase sealed drinking water, you should sanitize bottles with one teaspoon of unscented household bleach per one quart of water. Rinse thoroughly and then fill the bottles. Add two drops of unscented household bleach to each gallon, seal tightly, date, and store in a cool, dark place. Water purification tablets can also be purchased at most sporting goods stores. FEMA recommends replacing stored water every six months to ensure freshness.

- **Food.** You may safely ration food if necessary, except for children and pregnant women. Use canned foods, dry mixes, and other staples that do not require cooking, water, or special preparation, but do be aware that insects can invade non-airtight food stores such as boxes and plastic bags. Be sure to include a manually operated can opener—as well as a camp stove, if you have one—in your survival kit. Replace food items every six months by rotating them into your pantry for use and purchasing replacements.

   You might also consider including ready-to-eat canned meats, such as tuna and chicken; canned fruits and vegetables; canned or boxed juices, milk, and soup; high-energy foods, such as peanut butter, jelly, low-sodium crackers, granola bars, and trail mix; vitamins; foods for infants or persons on special diets; cookies and hard candy; instant coffee; and cereals and powdered milk. It only makes sense to try to store some of the same types of food that you and your children are used to eating. When a massive ice storm swept through the rural Midwest, causing electrical transformers to explode right and left, some areas were without electrical power for more than three weeks, and rural residents were thankful that they'd included some "regular" food in their emergency stores. FEMA has a downloadable document to aid you in assembling your readiness kit (www.fema.gov).

- **First aid kit.** Assemble a first aid kit for your home and for each of your vehicles. Each kit should include a first aid manual, adhesive bandages in assorted sizes, safety pins, cleansing agents (isopropyl alcohol, hydrogen peroxide), antibiotic ointment, at least two pairs of latex gloves, petroleum jelly, 2-inch and 4-inch sterile gauze pads (four to six of each size), three triangular bandages, three rolls each of 2-inch and 3-inch sterile roller bandages, cotton balls, scissors, tweezers, needles, moist towelettes, a thermometer, tongue depressors, and sunscreen. Also include an extra pair of prescription glasses or contact lenses as well as nonprescription drugs, such as pain relievers, antacids, and cough/cold medicines, in your disaster kit. If you require routine prescription medications, ask your physician or pharmacist about storing them properly.

- **Tools and other items.** Include a portable, battery-powered radio and flashlight, extra batteries, signal flares, matches in a waterproof container, a shut-off wrench, pliers, a shovel, a whistle, an ABC fire extinguisher, a tube tent, mess kits (or paper cups, plates, and plastic utensils), an all-purpose knife, household liquid bleach or water purification tablets to treat drinking water, sugar, salt, pepper, aluminum foil, plastic wrap, and resealable plastic bags.

- **Sanitation and hygiene items.** Gather a washcloth, a towel, soap, hand sanitizer, liquid detergent, toothpaste, toothbrushes, shampoo, deodorants, a comb and brush, a razor, shaving cream, lip balm, insect repellent, contact lens solutions, a mirror, feminine supplies, heavy-duty plastic garbage bags and ties, toilet paper, and a medium-sized plastic bucket with a tight lid.

- **Household documents and contact numbers.** Have copies of important household documents and contact numbers ready to go on short notice and stored in a watertight container. Include your personal identification; cash or traveler's checks and a credit card; copies of important documents (birth certificates, marriage certificate, driver's licenses, Social Security cards, passports, wills, deeds, inventory of household goods, insurance papers, immunization records, bank and credit card account numbers, and stocks and bonds); an emergency contact information list, a map of the area and information on safe places you can go; and an extra set of car keys and house keys.

- **Clothes and bedding.** Keep the following for each member of the household: a complete change of clothing and footwear, three sets each of socks and underwear, sturdy work shoes or boots, rain gear, a hat and gloves, sunglasses, a blanket or sleeping bag, and pillows.

Gardening: The Land

Whether you plan to garden commercially or just want to reduce your reliance on the grocery store for food, no farm feels quite complete without some kind of garden. The garden is a sacred place, loved by those who know it. It provides a respite from the hectic world, a chance to enjoy nature at its finest, and the opportunity to feast on a gift of the best food available anywhere. After all, who can resist the flavor of garden-fresh vegetables? Fruit so juicy that you have to wash your face, hands, and arms after eating it? Berries you can't stop eating, with a sumptuousness never found in the grocery store? Sweet corn so sugary that you know how it got its name? Carrots that are perfectly crunchy and potatoes that are naturally creamy? Greens so light and crisp that they don't taste like something your mother once forced you to eat? Squash that melts in your mouth?

Who is not awed by the miracle of new life pushing up from the soil after a warm spring rain just days or weeks after setting out seed? And who isn't cheered by spring-flowering bulbs poking up through the snow? Over the years, my own garden has changed in size, shape, and style. It will continue to change, as gardens do. At one time, we gardened on more than two acres and sold a good deal of our produce to friends and acquaintances. We thought about operating a community-supported agriculture (CSA) garden, but by then we were milking cows. We couldn't do both labor-intensive enterprises well, so we cut back on the size of the garden.

Now our garden is small, consisting of some outdoor beds and indoor containers that help keep us in fresh, organic produce all year. We don't grow all of our fruits and vegetables by any means, but we put a good

Your vegetable garden can be the size of a flowerbed or a field.

dent in our need to purchase produce. Our plans for the future include a year-round greenhouse just large enough to meet our needs. We want it to be attached to the house, but because that's an expensive step, it won't happen for a while. We treat it as part of the evolution of our current farm.

You, too, can grow crops on the field-scale size: acres instead of rows or beds. Some crops—such as field corn, soybeans, and hay—are the main enterprise, year-in and year-out, for many midsize and large commercial farms. Small and hobby farms often incorporate some field cropping, too; for example, they may grow several acres of pumpkins or a sweet corn patch, or they may put up hay for use on the farm. Field cropping and gardening follow many of the same practices, just on different scales. The principles and the underlying factors that predispose gardens of any size for success all count on the same things: soil, sun, water, and seeds.

Choosing the Site

Where you place a garden has a lot to do with its success. Traditionally, gardeners placed kitchen gardens close to—you guessed it—the kitchen door. If possible, it is still a good idea to place your garden in close proximity to the house. When it comes to gardens, the old adage "out of sight, out of mind" tends to apply. And it's amazing how quickly a crop of weeds can smother what you so bravely planted if you don't see it every day. A garden as near to the kitchen door as feasible is inviting, and if you pass it multiple times a day, you're more likely to keep weeds at bay and harvest your vegetables. Also, water for irrigation is usually readily available close to the house, whereas stringing a mile-long hose to water the tomatoes becomes a real chore.

To succeed in growing a wide variety of crops, your garden should receive at least six hours of sunlight a day, and more is even better. In shady patches, you can grow cool-season crops such as lettuce, broccoli, cauliflower, cabbage, carrots, and radishes, but warm-season plants such as tomatoes, peppers, corn, melons, and berries will not fare well at all. A level or gently sloping site with southern exposure is ideal for maximum sun exposure. On steep slopes, you will need to incorporate terracing: plant your tallest crops on the north end of the garden plot, working your way down to the shortest crops on the south end, which will keep tall plants from shading the shorter ones.

Wind can damage plants, especially young transplants, so selecting a site that is somewhat protected by a natural or man-made windbreak will increase the success of your garden. Although trees can provide a windbreak, they also create unwanted shade, and their roots can suck the water out of the soil almost as fast as you can put it in. Try to keep trees at least 50 to 60 feet from the edge of your garden.

Evaluating the Soil

Most people look at the ground and just see dirt, but, to a farmer, dirt is something greater. It is soil, and when combined with solar energy, air, and water, it forms the life force of the planet.

Check into local resources that can help you test your soil.

Types of Soil

Soil starts with minerals. Most of minerals were formed into rocks eons ago by geologic processes like volcanoes and earthquakes. Then, over more eons, the rocks were crushed, pulverized, and moved by erosion from wind and water, which carried them from mountains to plains and river bottoms. Forces like wind, sun, and rain break large rocks into smaller ones and small rocks into still smaller ones. Bacteria, lichens, algae, and insects get into the act, helping to further break down rocks. Chemicals in the soil, like acids and enzymes, keep the process going. The broken-down rocks combine with *humus*, which is organic matter from decaying plants and animals, to yield soil minerals.

You can generally classify soil minerals into three different categories based on size: *sand* is the largest particle, *silt* is intermediate, and *clay* is the smallest. The percentage of each category in a soil determines the soil's texture as well as its physical properties. An ideal soil texture consists of equal parts of each size category and is called *loam*. Most of the time, however, one size category dominates, so you end up with breakdowns like sandy loam or clay loam. The particle sizes affect how quickly soil drains and dries out. Soils high in sand drain and dry quickly, soils high in clay drain and dry slowly, and soils high in silt are somewhere in between.

Some soils contain little or no organic matter; they grow poor vegetation, if any at all. But healthy soil is vibrant and alive, containing living tissue in the form of roots, worms, insects, bacteria, and other microorganisms. Soil scientists estimate that a single tablespoon of good soil contains tens of millions of living organisms. These healthy soils grow abundant crops and native plants.

The Water Cycle

The water cycle is the movement of water between the atmosphere and the earth. Some water runs off the land to enter streams, rivers, lakes, and, ultimately, the ocean, to be evaporated into the atmosphere and rained down once again. But, in a healthy ecosystem with soil that is high in organic matter, much of the water that falls from the sky as precipitation is absorbed into the soil. Some of what is absorbed enters the groundwater, some evaporates directly from the soil to the atmosphere, and plants use the rest for transpiration. *Transpiration* is essentially the process by which plants obtain nutrients from the soil via water sucked in through the roots. Eventually, the excess water evaporates from the leaves of the plants.

Some folks are lucky enough to enjoy naturally fertile soils that are rich in humus and can grow bountiful crops with little effort—just throw some seeds in the ground and wait for harvest. But most of us are not so fortunate. In some cases, the soil was never rich in organic matter, which is my current situation in the Rocky Mountains, where the breakdown of rock hasn't progressed too far, and the long winters and dry conditions don't favor the quick conversion of dead plants, animals, and manure to humus. In other cases, the soil is worn out from bad practices like monocropping; applying too much or too many years' worth of chemical fertilizers, pesticides, and herbicides; or applying too much surface irrigation water that was high in salt. We need to take measures to improve our soil by adding amendments that increase humus in the soil, adjust the pH, increase and/or balance the nutrients, and improve the texture of the soil so that it is easier to work with and drains well while still retaining moisture. After choosing a site for a garden or field crop, it's a good idea to run some tests that will help you decide what amendments to add and when to add them.

Soil Tests

Although you can purchase home soil testing kits, I recommend checking with your extension agent to find out about labs that do the tests commercially in your state. Many state land-grant universities have their own testing labs that provide reasonably priced services to residents. Under most circumstances, you won't have to run tests more than once every three to five years, so the cost isn't too much of a burden. Lab test results are usually more accurate than those you get from a home

Soil tests help you find out what types of plants will grow best in your soil and what types of amendments you should add.

test kit. When pH results come back highly acidic or highly alkaline, purchasing a home test kit specifically for pH is a good idea so that you can monitor changes to the pH level as you adjust it with soil amendments.

Commercial labs (including those run by land-grant universities) will generally supply you with a container and paperwork to complete. Most ask you what crop you are growing because a technician will interpret the test results and make recommendations for soil amendments based on the type of crop.

Soil tests evaluate the pH of your soil, the major nutrients (e.g., nitrogen, phosphorous, and potassium), and the micronutrients (e.g., calcium, magnesium, sodium, and sulfur). The results can help you decide on appropriate plants for your soil, on soil preparation techniques, and on amounts and types of fertilizer to apply. They will also help you avoid overfertilization, which can stimulate excessive plant growth, increase the likelihood of some diseases, or cause pollution of your drinking water supply and your area's rivers and lakes.

Whether running a home sample or having a lab test your soil, the accuracy of your test results is highly dependent on the quality of the sample you take. Follow these steps:

- **Avoid contamination.** The best tool to use is a soil probe, which you can purchase through most garden-supply outlets. If you don't have access to one, use a stainless-steel or chrome-plated trowel or spade. Avoid using brass, bronze, or galvanized tools because these can contaminate the sample with copper, zinc, or both. Use a clean container to combine all samples; even small amounts of residual fertilizer, lime, or cleaning chemicals can give false results.

- **Take samples in different spots.** To fully represent the garden or field, you need to take anywhere from five to twenty samples from different spots. Sample throughout the rooting zone—the top 6–8 inches for gardens and crops other than hay, and the top 3–4 inches for hay fields.

- **Create soil composites.** Mix the samples from your garden or field together to create

Compost produces nutrient-rich soil in which plants can thrive.

a soil composite. If you are looking at distinct growing areas (say a garden, a hay field, and an orchard), take separate composites for each and have them analyzed separately. Mix the soil thoroughly.

Improving Soil Health

For most soils, the first step to improving soil health is to increase the organic matter, and adding compost, leaf litter (partially decomposed leaves), or well-aged manure is the best way to do that—particularly for garden-sized plots. From here on out, anytime I mention anything about adding compost to the garden, you can substitute leaf litter or well-aged manure in the application.

Compost

Compost is like black gold. Furthermore, by composting organic waste, you reduce the amount of trash going to landfills and incinerators, and you feed the plants around your home. For urban and suburban dwellers, a compost tumbler or bin is a great way to do something good for the planet and for yourself.

Composting is a natural biological process carried out by organisms such as bacteria, fungi, insects, and earthworms. The organisms responsible for composting consume organic materials and oxygen in order to grow and reproduce. They produce carbon dioxide, water vapor, and heat as by-products of their activities. From start to finish, the composting materials transform from a diverse mixture of individual ingredients, such as grass clippings, leaves, food waste, and manure, to a uniform soil-like material. Six key elements affect the process.

- **Carbon-to-nitrogen ratio:** Carbon materials are the "browns" that go into the compost pile. They include dead leaves, twigs, cornstalks, hay, straw, dried grass clippings, sawdust, wood chips, finished compost, and shredded newspaper. They provide bulk and air gaps, and they help keep the pile from going sour. Nitrogen materials are the "greens." They include fresh grass clippings, fruits and vegetables, coffee grounds, eggshells, garden waste, and manure. A well-balanced proportion of carbon and nitrogen ensures a good supply of all nutrients and allows composting to proceed rapidly. The ideal ratio is about three parts carbon to one part nitrogen by volume. In other words, if you place a layer of browns that is about 3 inches deep on the bottom of your pile, the next layer of greens can be about 1 inch deep.

- **Surface area:** The smaller the materials that go into the pile, the faster they will compost. Smaller chunks give more surface area for organisms to work on, so break up those branches and shred that newspaper.
- **Aeration:** Aerobic organisms carry on the process of composting. You need good airflow throughout the pile to supply these workers with the oxygen they need to breathe. Aerating and turning the pile from time to time is important to allow the composting process to continue. Use a potato fork (similar to a pitchfork, but with a shorter handle and sturdier tines) or some other tool that allows you to penetrate to the bottom of the pile, lift the composting materials, and turn them over. As you will see, the middle of the pile, which is where the heat will be highest, tends to compost fastest.

We turn our compost in spring, as soon as it thaws, and then once a month or so during the summer and fall. At each turning, we take some finished compost off the bottom of pile. For small operations that are primarily dealing with household waste and a small amount of yard or garden waste, commercial compost tumblers work well. For a midsize operation, running several bins or piles works well. Completely filling one bin before starting on a second, rather than randomly adding to whichever one is closest, will yield a uniformly composted result; the latter method will set the process back somewhat. Large operations, like those with lots of manure, typically build free-form piles and turn them with a tractor fitted with a front-loading bucket. Medium and large piles also benefit from several perforated plastic pipes inserted vertically into the pile to increase airflow.

Vegetable matter and eggshells are among the nitrogen-producing "greens" in the compost pile.

- **Moisture:** Your pile needs to be damp most of the time, but not sopping wet. A dry pile leaves the organisms thirsty, but a too-wet pile loses critical air space that allows the organisms to breathe. During dry spells, remember to add water once or more each week because the composting process is very slow when materials are dry.
- **Temperature:** The organisms generate heat while they work, thus raising the temperature of the composting materials. The ideal temperature is in a range from about 135 to 160 degrees Fahrenheit, although temperatures in a pile can sometimes exceed 160 degrees.
- **Time:** It can take anywhere from several weeks to more than a year to produce compost,

depending on the ingredients and conditions in a pile. Compost is typically ready for use in three to four months during warm weather, given regular turning, adequate moisture, and a good mixture of materials. Methods that receive little or no turning usually require more than a year to produce compost that is ready to use.

You can easily build compost piles by alternating layers of green and brown material, keeping in mind the ideal ratio of about three parts carbon to one part nitrogen by volume. We pick up some night crawlers from a bait-supply store each spring and dump them on our compost piles to help speed up activity.

Manure and Fibrous Plant Matter

If you have livestock, you have a readily available supply of fertilizer for your compost pile and your land. Never use fresh manure in the garden during the growing season, however, because it can burn plants due to high nitrogen content, and it may contain viable weed seeds or pathogens. However, if you work some fresh manure into the garden during the late fall, by spring it will be broken down sufficiently and will not cause problems. Applying a thin layer of fresh manure to farm fields in the spring, prior to tilling and planting, is a time-tested method of enriching your land that is best accomplished with a manure spreader. The spreader lays manure out in well-broken pieces that sun and precipitation quickly break down. For fields in which you overwinter cattle, dragging a harrow over the ground will achieve much the same effect of breaking up cow piles into small pieces and distributing them more evenly.

Another approach to improving soil health is to grow a "green manure" crop. Green manures are plants grown for the sole purpose of being turned back into the soil in order to enrich it. Crops like clover, field peas, oats, annual rye, wheat (spring or winter), and buckwheat are common green manures. When you use legumes (plants that pull nitrogen from the atmosphere and affix it to their

A green manure crop being turned back into the soil.

Fertilizer often comes in the form of small pellets that you mix into the soil.

roots; e.g., clover, alfalfa, and field peas) as green manures, you get an extra boost of free nitrogen fertilizer. Whereas you can add compost, leaf litter, and well-aged manure to your soil immediately before planting, green manures need to break down somewhat after you turn them under but before you plant your crop because the soil needs time to compost the fresh material. After you turn down the green manure, wait three to four weeks before planting your next crop. If you have a site where the soil is really worn out, planting several successive green manure crops is the best way to bring it back.

The final approach to improving your soil's organic matter is to turn fibrous plant matter, such as sawdust, chopped straw, or shredded newspaper, into the soil. However, because of their high carbon concentration, these materials are slow to break down, and they can steal nutrients such as nitrogen from your crop in the process. If you go this route, add additional sources of nitrogen to help speed up the process without depleting your soil.

Fertilizers and pH Adjustment

Although compost, manure, green manure, and fibrous materials add some nutrients as well as organic matter, your soil may require additional amendments to feed the plants that you plan to grow. There are two approaches. The first is to feed the soil and let it take care of the plants, and the second, known as *foliar feeding*, is to feed the plants directly with soluble feeds that they can take up through their roots and into their leaves.

I believe in feeding the soil and letting it take care of plants through the natural system that Mother Nature developed. However, if you are starting out with soil that is far out of balance, foliar feeding sometimes can be an excellent short-term fix to help your crops along while you rebuild the soil. If foliar feeding is necessary, you may need to do it as frequently as six to ten times during the growing season, starting with seedling emergence and continuing until just before harvest. Apply foliar feeds in the early morning or early evening, rather than during the heat of the day, so that you lose fewer nutrients to plant transpiration. You can brew compost tea and use it as a foliar feed, or you can purchase fertilizers that are appropriate for foliar feeding.

The letters "pH" refer to the percentage of free hydrogen ions; "pH value" describes the acidity or alkalinity of a substance on a scale from 0 to 14. If a substance has a pH value below 7, it is acidic; if it is above 7, it is alkaline; and 7 itself is neutral. The pH scale is interesting because, as you move up and down the scale, changes in pH are expressed in factors of 10. For example, a pH of 6 is 10 times more acidic than a pH of 7. A pH of 5 is 10 times more acidic than a pH of 6 and 100 times more acidic than a pH of 7. A pH of 10 is 10 times more alkaline than a pH of 9, 100 times more alkaline than a pH of 8, and 1,000 times more alkaline than a pH of 7.

Most of the plants we want to grow do best in soils with a pH between 6.0 and 6.8—which is slightly acidic—although certain plants (such as azaleas, blueberries, and cranberries) prefer more acidic soils, and others (such as peppers, yams, and legumes) appreciate slightly alkaline soils. In the United States, most soils have a pH level between 4 and 9.

A crop of young bell peppers. This vegetable tends to do best in slightly alkaline soil.

You can bring up acidic soil by adding limestone. You can bring down alkaline soil by using sulfur or aluminum sulfate. It is best to adjust pH in several smaller applications of limestone or sulfur throughout the year versus one big application. Again, running a soil test is the quickest way to ascertain what your soil needs.

Bagged fertilizer generally contains three elements in various amounts, indicated by a series of three numbers on the package. A common type is 10-10-10. The first number in the series indicates the nitrogen content. Nitrogen is essential to the growth of leaves. Indicators that your plants have too little nitrogen are yellowing leaves and very little growth. Nitrogen is essential, yes, but if you go overboard, you'll end up with a lot of healthy leaves but not much fruit.

The second number represents the phosphorous/phosphate content. This stimulates root growth, helps plants mature and fruit develop, and remains in the soil longer than nitrogen. Without phosphorus, plants grow slowly and may be stunted.

The third number stands for the potassium content; without potassium, plants will be weak stemmed and slow growing. Potassium helps plants resist disease and cope with drought, heat, and cold. Determining what your soil needs based on the results of a soil test is the quickest and most accurate way to decide which fertilizer balance to go with.

Both organic and chemical fertilizers exist, and both will boost your plants' growth and health. Organic compounds enrich the soil health in ways chemicals do not. They also dole out nutrients at a slower rate, lowering the chance of damaging your plants. Chemical fertilizers have the advantage of lower cost (usually), are easier to transport, distribute nutrients more evenly, and are readily available for plant use.

Preparing the Land

Whether growing crops in a field or planting a garden, there are some steps required to prepare the seedbed for your crop. Farmers use the term "tillage" for the process of preparing the seedbed. For smaller gardens, I prefer the deep-bed method to row plantings with traditional tillage. However,

Raised-bed gardening was used extensively in 19th-century Paris and is still a popular planting method today.

if you have the space, the tractor-driven equipment, and the urge to spend some time driving the tractor, row planting definitely works. There's no one "right" way to garden, so don't be afraid to mix and match methods—there are master gardeners who espouse different techniques and layouts. Research, ponder, and glean what you think will work well for you, then try it out. Adjust. Upgrade. Every step of the journey is the journey!

Deep Beds

Many people credit an Englishman named Alan Chadwick with pioneering the deep-bed method in the United States as early as the 1930s, when he began experimenting with the practice at the University of California at Santa Cruz. The method is based on the traditional raised beds of French market gardeners. During the nineteenth century, Parisian gardeners were able to grow more than 100 pounds of produce annually for every person in the city. They did this through the use of beds that were dug deeply and then built up to heights of 18 inches above the paths that bordered them, adding plenty of well-aged horse manure (something that was quite abundant at the time). By spacing plants closely, they could maximize the production in a given area, and by using glass *cloches* (bell-shaped mini-greenhouses that sat over individual plants or clumps of plants), they were able to produce throughout the winter.

The ideal width for a bed is about 3 feet; you want to be able to straddle it. Although beds can be any length you want them to be, I prefer relatively short beds (10 to 20 feet long) with cross paths, which make moving around the garden convenient. I make most paths wide enough to accommodate a two-wheeled wheelbarrow or a garden wagon. Mulching paths with 3 inches or so of straw, wood chips, or shredded newspaper helps control weeds and maintain moisture in the beds while at the same time reducing mud and mess. Try to never walk on the beds themselves because this compacts the soil, and the success of the beds relies in part on the airy soil structure you create.

Another proven method of planting is to arrange your crops in blocks instead of rows, using an equidistant spacing pattern to create a nearly solid leaf canopy or "living mulch." This reduces weed pressure and conserves moisture. In a similar fashion, "wide row gardening" utilizes the plants themselves as mulch and yields much more produce per square foot than does the traditional row method of one plant width.

Vegetable and herb gardens have traditionally been oblong, because oblong plots are easier to work with a tractor. But when using beds, you can design your beds in fanciful patterns. In recent years, I have been developing a series of beds in unusual shapes that follow the natural lay of the land. In the beds, I intermingle flowers, herbs, and vegetables. They add a landscaped feel to the area around the house while producing food.

Double-Digging Beds

Some gardeners advocate a method called double-digging. To double-dig your beds, follow these steps:

1. Starting at one end of the bed, dig out a layer (about the depth of the spade) in a trench that runs the width of the bed. Set this aside in your wheelbarrow or garden cart.
2. Loosen the soil at the bottom of the trench by digging it with a fork and turning it over.
3. Add compost or manure to the bottom of the trench so that it reaches about halfway to the top of the trench. Add any other slow-release amendments.
4. Dig another trench adjacent to the first, placing the soil from the second trench on top of your first trench.
5. Repeat steps 2 through 4 until you have dug and treated the entire bed.
6. Use the soil from the first trench to cover the last trench.
7. Shape the bed to mound it slightly in the middle.

Tool Tips

For small gardens, you'll need a flat spade and a pointed spade for digging. A good digging fork is a multipurpose tool that you can use to break up and turn soil in the garden, harvest potatoes, and clean up manure around the barn. My favorite hoe is the collinear hoe designed by gardening guru Eliot Coleman and available from Johnny's Selected Seeds (see Resources). This style of hoe takes the "backbreaking" out of controlling weeds because it allows you to stand up straight while you work. Another innovative type of hoe is the "stirrup" hoe, which allows you to push it as well as pull it.

Rear-tine rototillers are easier to walk behind than those with tines in front, and most brands of rear-tine models do a better job breaking up a deeper layer of dirt and pulverizing the soil particles. For medium-size gardens (say, more than a couple of beds but less than an acre or two), a walking tractor can be a cost-effective alternative to a compact tractor or utility vehicle with implements. A walking tractor is like a big, heavy-duty rotary tiller, but the tiller isn't permanently mounted; instead, it is one of several different attachments that you can add, such as seeders, hillers, and mowers.

A word to the wise here: Seeds of all types can travel through a horse's digestive system and still be viable, which is important to know when it comes to using horse manure as garden fertilizer. Yes, horse droppings contain nitrogen, but it's best to compost them first or use them to create compost tea in order to avoid seeding your garden with a host of grass and weed seeds.

A green onion crop planted in typical rows.

Lazy Person's Bed

Personally, I find double-digging to be too much work. My method of preparing a bed does require one overwintering period, so it has a disadvantage if you are in a hurry; but if you plan ahead, it is quite easy. First, in late summer or early fall, mark the area where you want to create your bed with rocks or stakes at the corners. Using your round-pointed shovel, turn over the top 4–8 inches of soil (depending on how easy it is to dig and turn) to put the existing vegetation down under the soil surface. As you're working, remove any large rocks from this layer. Next, top-dress the bed generously with 3–6 inches of compost (this is one time you can substitute fresh manure if you don't have compost or aged manure). Last, sprinkle a layer of straw mulch lightly over the bed (1–2 inches is good). Now, sit back and wait.

In late winter, head out with your spade and dig around, mixing up the soil, compost, and straw as well as you can to about the depth of the spade. If you live where soil freezes, perform this step as soon as the soil thaws enough to work it; in less severe climates, a nice afternoon in early February would probably be the time to do it. If you are adding any slow-release amendments, such as wood ash, greensand, lime, or sulfur, put them on top before you do your mixing. Add another inch or so of compost to the top of the bed and then take another break until the weather warms up enough to start thinking about planting.

As planting time approaches in late spring, remix the bed lightly to incorporate the last top dressing. Slightly mound the bed in the middle with a rake. Each fall, you can add compost and some mulch, and each spring, you can turn it under. Over several years, your beds will improve, and you will see the difference in your crops.

Field Tillage

Traditional tillage involves plowing and harrowing. Plowing breaks up the soil surface, turns under vegetation and crop residue, and incorporates manure, lime, or fertilizer. Several types of plows are available. The *moldboard* plow, which earned the name "prairie buster" as settlers opened the frontier in the nineteenth century, can turn over up to 2 feet of soil with sod or dense vegetation on it. *Chisel* plows are narrower and don't dig as deeply, making them inadequate for plowing a field for the first time. Their advantage comes into play on fields that have previously been planted in annual crops. Chisel plows disturb less soil, causing less compaction of subsoil layers. Chiseled land is less erosive.

When plowing land that has been farmed for annual crops in the recent past, a compact tractor may do the job, but if you are plowing land that has been in sod for years, you will need at least a utility-size tractor. (See Chapter 3 for a discussion of tractors and implements.)

A freshly plowed field.

Harrowing follows plowing and is intended to create a smooth soil surface by breaking up the furrows and clods of dirt thrown up by the plow—it's comparable to giving a garden bed a final raking. It prepares a fine and even seedbed that creates the ideal environment for germination. Harrowing is often a multistep process, depending on the type and depth of plowing and the crop for which the field is being prepared; for example, very fine seeds require a more even and finely prepared seedbed than do larger seeds. A disk harrow is often the first tool used following the plow because it is capable of quickly breaking down the rough surface and clods of soil left by breaking the ground. For a field that doesn't require a very smooth seedbed, you can often do the work in a single pass by using a disk harrow with a set of chains following it, but for a very smooth field, you should make a second pass with a spring-tooth harrow.

Try to minimize the amount of tillage you do because every pass of a heavy machine ultimately compacts soil and reduces *tilth* (soil health). Some farmers will only plant into perfectly smooth seedbeds, but many crops—particularly those with large seeds—don't always need that level of perfection to perform well.

Tillage Tips

Never perform any tillage operations in fields or gardens when the soil is wet and muddy. Doing so causes soil to become compacted, which blocks root penetration and results in more *sheet runoff* (water running along the surface of the ground rather than being absorbed by the soil) and erosion. It also can bog your tiller to the hubs, and cleaning up that mess is no fun.

Gardening: The Planting

Seeds are simply miraculous. They are bundles of life—plants in embryonic form—just waiting for the right conditions to grow and reproduce.

Plant Basics

You can group plants into three primary classes according to their life cycles. Some plants are annuals. They must go through their whole life cycle, including producing seeds, during the year in which they sprout. Biennials are plants that take two years to complete a growth cycle and yield seeds. Perennial plants live from year to year; shrubs and trees are all perennials. Some annual plants are self-seeding, so, from a gardener's point of view, they behave like perennials because you don't have to replant them yourself.

Annuals reproduce strictly from seeds, so scientists consider their reproduction sexual (needing to be pollinated to form seeds). Biennials and perennials can reproduce sexually from seeds, but these plants may also reproduce asexually from existing roots.

Plant Families

Plants, like people, belong to families. Like human families, plant families share significant traits. Some of the important families that farmers and gardeners deal with are:

- **Apiaceae (formerly Umbelliferae):** Carrots are probably the most common member of this family in gardens, but you may also plant celeriac or celery. Apiaceae herbs include anise, coriander, cumin, dill, fennel, and parsley.

A young cucumber and a cucumber blossom, both of which are edible, on a plant.

- **Chenopodiaceae:** Beets and spinach are the two most common members of Chenopodiaceae, but quinoa, Swiss chard, and sea kale are other members you might consider planting. The "seeds" that you plant are actually tiny fruits containing the real seeds inside them, so when you plant one "seed," multiple plants come up. This means that you have to thin your plantings. An advantage of planting Chenopodiaceae is that the members of this family have a dense root system that breaks up compacted soil. Grow beets for livestock feed as well as for humans.

- **Compositae:** Lettuce is probably the most famous member of the Compositae family. Included as well are artichokes, cardoons, chicory, endive, salsify, and scorzonera. Unfortunately, this family also brings us some weedy members—such as dandelions and thistles—that find their way into gardens, lawns, and fields against our wishes.

- **Cucurbitaceae:** Yum, yum, yum. This family brings us cucumbers, melons, pumpkins, and squash. All members of this family are warm-season crops, so do not place seeds or transplants in the ground until the danger of frost has passed. Although generally grown for their fruit, Cucurbitaceae plants have edible blossoms, and you can add them to salads.

- **Cruciferae (or Brassicaceae):** Also called the cabbage family, the Cruciferae family includes broccoli, Brussels sprouts, common and Chinese cabbage, cresses, kale, kohlrabi, radishes, rutabagas, and turnips. All of the plants in this family are biennials. They have a two-year growth cycle to create seeds, although for garden produce, you pick the crop during its first year. They are cool-season crops, so they can go into the garden early, producing some of the first goodies to harvest; as heat comes on, they wane. Plant Brussels sprouts in summer for harvest in the fall because frost sweetens the result.

- **Lamiaceae:** Known as the mint family, this group brings us a wide variety of herbs, ranging from basil and bee balm to rosemary, sage, and thyme. These are highly aromatic herbs that are widely used in cooking and sometimes used for their medicinal properties.

- **Lauraceae:** This is a family that doesn't rest on its laurels. It brings us a number of perennials, including avocados, as well as bay leaf and cinnamon, two very popular herbs.

- **Leguminosae:** Most living organisms use the ammonia form of nitrogen (NH_3) to build the protein that is required for body functions. The atmosphere consists of about 80 percent nitrogen; however, this nitrogen is available in a nonusable form of gas (N_2). Legumes are unique in that they are able to "fix" that atmospheric form of nitrogen in their roots, changing it to the usable NH3 form. By doing this, they provide part of their own fertilizer while increasing soil fertility, which other plants can then take advantage of. Legumes accomplish this feat thanks to a mutually beneficial relationship with a group of soil microorganisms known as rhizobia. Rhizobia live in nodules on the roots of the legumes. The plant contributes energy, in the form of sugars and other nutritional factors, for the bacteria so that the bacteria can provide the plants with the required nitrogen in a form it can use. To enhance nitrogen fixing, you can inoculate seeds with appropriate rhizobia before planting.Legumes are also special because they provide more dietary protein for animals (including us two-legged creatures) than do any other plants. The legumes include such garden favorites as peas, beans, lentils, and peanuts. Crops for livestock feed include field peas, alfalfa, clover, vetches, and birdsfoot trefoil.

- **Liliaceae:** This is one of my favorite families. Its members include asparagus, chives, garlic, leeks, onions, and scallions. This family has flowering members as well, such as daylilies, crocuses, daffodils, hostas, hyacinths, lilies-of-the-valley, and tulips. The aloe plant, known for its medicinal properties, is also a member of the clan.

- **Poaceae (formerly Gramineae):** Better known as the grass family, Poaceae is represented in the garden by only one family member—sweet corn. Outside of the garden environment, Poaceae includes many cereal grains, such as oats, rice, rye, and wheat, and forage crops for the pasture, such as bromegrass, orchard grass, and timothy.

- **Polygonceae:** Rhubarb and sorrel are the two members of this family found in the garden, but the most famous member is buckwheat, which is a major green-manure crop as well as a minor grain crop. An added bonus: Buckwheat hulls can be used as stuffing for pillows if you have leftover seed you want to use up.

- **Rosaceae:** As the name implies, roses are members of this family, but also included are many of our favorite fruits: almonds, apricots, apples, blackberries, cherries, nectarines, peaches, pears, plums, raspberries, and strawberries. All members of this family are dependent on insects (primarily bees) for pollination.

- **Rutaceae:** Those of you who live in southern-tier states and some parts of California are probably familiar with this family, which supports a large amount of economic activity in those regions. Rutaceae members include the citrus fruits: grapefruit, lemons, limes, oranges, and tangerines. For those relegated to northern climes, if you are dedicated, you may get citrus fruit from trees grown indoors in pots.

- **Soanaceae:** This group includes some all-time favorites—eggplant, peppers, potatoes, and tomatoes. Soanaceae also includes some dark horses, such as belladonna, deadly nightshade, and

The popular strawberry is classified in the same plant family as some other popular fruits as well as the rose.

Members of the Rutaceae family, including citrus fruits, need the warmer weather of the southern states.

tobacco. Ground cherries, jimson weed, and petunias are members of the tribe as well. The edible members of this clan are good sources of vitamin C. Soanaceae plants like slightly acidic soil, so you don't want to add lime where you will plant them. They like potassium, too, so adding a source of potash, such as wood ash, is a good idea.

- **Vitaceae:** The final plant family of importance for gardeners, farmers, and wine drinkers is the Vitaceae family, which brings us the grape. Most areas of the country can grow one or more varieties of grape, but grapes are very geographically specific. Some grapes can't grow in high-humidity areas, whereas others crave humidity. Some tolerate heat well; others are hardy enough to survive a Minnesota winter. If you are interested in trying grapes, research varieties suitable to your area before purchasing any vines.

Hardiness

You can successfully grow most annuals throughout the United States. In cold climates, growing seasons are shorter, so you may need to start less hardy annuals indoors and transplant them outdoors if they are to reach maturity before frost nips them. Unfortunately, not all biennials and perennials grow everywhere; we have climate to thank for this harsh fact.

In 1960, Henry T. Skinner, then-director of the US National Arboretum, in cooperation with the American Horticultural Society, supervised the creation of the Plant Hardiness Zone Map. The map is broken down into hardiness regions based primarily on average annual minimum winter temperatures. It has become a valuable tool for gardeners interested in growing perennials because it can be discouraging and expensive to find out by trial and error that certain trees, bushes, or flowers cannot survive your particular winter temperatures.

Outside of Your Zone

If you just love certain plants too much to abide by your growing zone, try starting them indoors or growing them indoors; for example, some citrus cultivars grown in large containers do produce if they receive enough light. In the case of some biennials, you'll have to find contentment in purchasing them, enjoying their blooming season in the garden spot of your choice, and replanting them each spring.

When shopping for perennials, you will see references to the zones in which the plants will grow. Plants may survive in warmer or colder zones than those listed, but a plant's range represents the optimum for that plant's performance. But be aware of *microclimates*, localized areas where weather conditions may vary from the norm. For example, a sheltered yard may regulate temperature extremes enough to support plants not normally adapted to the region, or a north-facing slope may be significantly cooler or windier than surrounding areas, thereby reducing the survival of plants normally adapted to the area.

Methods of Propagation

Plants that grow from seeds are classified as either *monocotyledons* (also known as monocots) or *dicotyledons* (dicots) depending on how their first leaves grow. Monocots, including many grasses and grains, start life with a single straight leaf. Most fruits, vegetables, and legumes are dicots, sending up two leaves. The dicots have

Grasses grow by sending up a single straight leaf.

thicker, tap-type roots, and their leaves are often patterned with veins in a weblike design. Seed coats protect most seeds from the elements until the time is right for them to grow. Some seeds only remain viable for a short period, measured in days, but others can remain dormant for many years and then spring to life when the time is right. Lotus seeds, for example, remain viable for thousands of years. Most garden seeds, however, last no more than a few years. Storing unused seeds in airtight containers in the refrigerator or freezer will help extend their life and, in some cases, cause them to sprout more easily.

When it comes to many heat-loving plants, transplants give you a head start on the growing season. They are readily available in the spring at garden centers, grocery stores, and discount stores. Starting with transplants is cost-effective as well.

Alternatively, you can start your own plants in the house with little effort. The biggest benefit of starting your own plants is that you have more seed varieties to choose from. Plus, it's almost essential if you want fully organic produce because commercially available transplants are often fertilized with petroleum-based fertilizers and treated with chemicals to protect them from insects and diseases.

Starting with Seeds

You can start seeds directly in the garden, indoors for transplanting outside, or in containers to remain inside. Seeds sprout when temperature, moisture, light, and other conditions are just perfect. For some seeds with particularly hard seed coats (usually indicated on the seed packet), breaking this *external dormancy* naturally may take years, but you can opt for some intervention strategies that trick the seed into coming to life when you want it to.

According to the North Carolina State University's Department of Horticultural Science, *scarification*—the process of breaking the seed coat by some means—enables moisture and gases to

Fruits and Vegetables

Common Name	Companions	Antagonists	Sowing
Asparagus	Tomatoes, parsley, basil		From roots in trenches; seeds do not yield for at least three years
Beans	Potatoes, marigolds, cucumbers, corn, strawberries, celery	Onions, garlic, gladiolus	Sow at least 2 inches deep in the garden
Beets	Lettuce, cabbage, onions	Beans, mustard	Sow in place in batches three weeks apart; presoak seed
Broccoli	Nasturtium, aromatic herbs,* potatoes, celery, dill, beet, onions	Strawberries, tomatoes, beans	Indoors four weeks before you want to set out or outdoors for later crops
Brussels sprouts	Nasturtium, oregano, potatoes, celery, dill, beets, onions	Strawberries, tomatoes, beans	Indoors (four weeks)
Cabbage	Nasturtium, oregano, potatoes, celery, dill, beets, onions	Strawberries, tomatoes, beans	Indoors (four weeks)
Carrots	Peas, lettuce, chives, onions, rosemary, sage, tomatoes	Dill	Sow in place, but not in very wet soil
Cauliflower	Nasturtium, oregano, potatoes, celery, dill, beets, onions	Strawberries, tomatoes, beans	Indoors (eight weeks)
Celery	Leeks, tomatoes, beans, cauliflower, cabbage		Indoors (four weeks); sowing in a cold frame is advantageous to get it going; plant in trenches
Cucumbers	Beans, nasturtium, corn, peas, radishes, sunflowers	Potatoes, aromatic herbs	Indoors (eight weeks) or under a cold frame
Corn	Potatoes, peas, beans, cucumbers, pumpkins, squash, marigolds, sunflowers		Outdoors after last frost; speed things up a little by starting in a tunnel
Eggplant	Beans		Indoors (eight to ten weeks) or outdoors in a cold frame
Garlic	Beets, strawberries, tomatoes, lettuce, chamomile	Peas, beans	Cloves in fall for harvest the next fall
Lettuce	Carrots, radishes, strawberries, cucumbers		Outdoors

* Aromatic herbs include basil, oregano, and rosemary

Fruits and Vegetables

Common Name	Companions	Antagonists	Sowing
Melons	Pumpkins, corn, nasturtium, radishes		Indoors (four to eight weeks) or outdoors in regions with a long growing season
Onions	Broccoli, marigolds, beets, carrots, eggplant, strawberries, tomatoes, lettuce	Peas and beans	Indoors (eight weeks) or outdoors for seeds, outdoors for sets
Peas	Potatoes, aromatic herbs, carrots, turnips, radishes, cucumbers, corn, beans	Onions, garlic, gladiolus	Outdoors at least 2 inches deep; presoak seeds
Peppers	Basil		Indoors (eight weeks) or outdoors in regions with a long growing season
Potatoes	Eggplant, aromatic herbs, beans, corn, cabbage, marigolds	Pumpkins, squash, cucumbers, sunflowers, tomatoes, raspberries	Sets in place after last frost or earlier in a cold frame; plant about 5 inches deep
Pumpkins	Corn	Potatoes	Indoors (four to eight weeks) or outdoors in regions with a long growing season
Radishes	Peas, nasturtium, lettuce, cucumbers, beets, spinach, melons, tomatoes, beans	Cabbage family	Outdoors; sow successive batches about two weeks apart for a constant supply
Shallots	Broccoli, marigolds, beets, carrots, eggplant, strawberries, tomatoes, lettuce	Peas, beans	Sets in place
Spinach	Strawberries, radishes		Indoors for an early start or outdoors about 1 inch deep
Squash (summer)	Nasturtium, corn, radishes		Indoors (three weeks)
Squash (winter)	Nasturtium, corn, radishes		Indoors (three weeks) or outdoors
Strawberries	Beans, spinach, lettuce	Cabbage family	Roots
Tomatoes	Chives, onions, parsley, asparagus, marigolds, nasturtium, carrots	Potatoes, cabbage family	Indoors (eight weeks)
Turnips	Peas, vetches, beans		Indoors (four weeks) or outdoors
Watermelons	Potatoes		Indoors (eight weeks) or outdoors in regions with a long growing season

A gardener plants a bean seed directly in the soil; no presprouting is necessary.

penetrate the impermeable surface. In nature, seed coats break as a result of freezing and thawing and of living creatures (ranging from microscopic organisms to birds and animals) working over the seeds. The easiest way to scarify seeds is to scratch the seed coat; an emery board works well.

A second type of dormancy is *internal dormancy*, which requires a period of "after-ripening" once the seed has reached maturity but before planting. Those species with shallow internal dormancy may simply require some time in dry storage. Seeds with deep internal dormancy require exposure to cold and warm temperatures (processes called *cold stratification* and *warm stratification*). In nature, seeds would naturally encounter changes in temperature to help bring them out of dormancy. To mimic this process in the comfort of your own home, perform cold stratification by mixing seeds with an equal amount of moist sand or planting medium in a closed container and storing the seeds and media in a refrigerator at approximately 40 degrees Fahrenheit. The length of storage time necessary varies from species to species. Warm stratification is essentially the same process, but with temperatures maintained at 68–86 degrees. The North Carolina State Horticulture Department also mentions "double dormancy," referring to seeds that must first be scarified and then go through stratification in order to germinate in appropriate climatic conditions.

After germination occurs, seedlings require ten to sixteen hours of intense light per day. Without strong light, the seedlings become spindly or *leggy* and may lean toward whatever light source is available. For indoor sowing, light from a south-facing window may be sufficient; otherwise, use a grow light for supplemental light.

Outdoors, you can scatter seed in the garden by hand or with a seeder that you push. Different models use different approaches for delivering the seed. Fluid, belt, and vacuum seeders are expensive specialty seeders, probably only justified for very large commercial gardens. Plate seeders, such as those made by Earthway Products of Indiana and the Lambert Corporation of Ohio, are reasonably priced and accurate in seed placement, making them an ideal choice for smaller gardens. Adding seed to a pasture to improve the diversity of grasses and legumes is easy to do with a broadcast seeder. Broadcast seeders are available as bags that fit over your shoulder, in larger units that you push in front of you, or in even larger setups that you pull behind a tractor or ATV. Seeding field crops such as corn and beans requires specialized planters or *seed drills*.

Transplants

You can start seedlings indoors in almost any kind of container, or you can use a soil blocker, which presses out soil cubes in sizes ranging from ¾ inch to 4 inches. I have used both containers and soil blockers, and, for large-scale planting, the blocks can't be beat. For starting a small batch of transplants, however, pots, yogurt cups, milk cartons, egg cartons, flats, or any container in which you can poke a hole in the bottom works fine. I keep plantings watered from beneath as much as possible, and, until the first seedlings emerge, I keep the containers or blocks lightly covered with plastic wrap (plastic produce bags or the bags you buy for cooking turkeys work well for containers, which can be set right in the bags). Make a few slits in the plastic to allow airflow.

Young marigolds started in containers for transport to an outdoor garden.

Some people recommend using only commercial potting mixes for starting seeds, but I like to use a home brew that incorporates my ever-improving garden soil, compost and/or well-aged manure, some commercially purchased sphagnum moss, and perlite. My proportions are four parts garden soil, four parts compost/manure, four parts moss, and one part perlite (for clay, add two parts coarse sand). This mixture works well in any container, and it does well in blocks if you keep stones and rocks from the garden out of it. A few grass seedlings or weeds may pop up in the transplants, but they are easy to pluck right out.

To water plants from beneath, run a piece of undyed wool yarn up through the bottom of the container and through the soil. The yarn will act as a wick that brings water up from the pan underneath the pot.

Layering, Cutting, and Division

Layering, cutting, and division are techniques that offer an alternative to starting plants from seeds when you wish to propagate perennial and biennial plants from a parent plant's stems or roots. Layering is suitable for many perennials with flexible stems/branches. It is the simplest and most reliable method to increase perennials. The idea is to produce roots on a stem while the stem is still attached to the parent plant. Select a healthy branch that is growing close to the ground and that is flexible enough to bend down to the soil. While holding the branch horizontally and close to the soil, bend the top 6– 10 inches of the branch into a vertical position. It's helpful to scrape the bark on the underside of the branch at the bend. Bury the bent, scraped portion 3–6 inches deep and anchor it with a wire loop. Insert a small stake to hold the top upright. Keep the branch moist. Once the root is established, cut the new plant from the parent plant.

You can start new growth from cuttings of established plants.

You can take cuttings any time during the late spring and summer from healthy, well-established plants (those taken in fall may need to overwinter in pots or temporary beds, so they take longer to fully develop). Healthy tip growth makes the best cuttings. Cut below a node to form a cutting that is 3–5 inches long and plant it in potting soil. Keep the soil well moistened and in partial shade until the roots form. Rooting can take anywhere from two weeks to two months to take (or *strike*), depending on the plant; the woodier the cutting, the longer it will take. Transplant once the roots take off.

Division is useful for multiplying healthy, established plants that may be two to four years old. Dig up an old plant and cut or pull it apart into sections. This works especially well for plants with tuber-type roots, such as irises. Replant the sections and keep them moist until the new plants are established.

Gardens, Groves, and Fields

Mention gardening, and the first thought that comes to mind for many people is a small vegetable patch or some ornamentals planted near the house. We farmers, however, can grow things on different scales and in different places. A field of alfalfa, just sprouting after a spring rain, or a grove of trees, grown for fruit, nuts, shade, or other forest products, are as much a farmer's garden as the rose bushes next to the house or the potatoes by the kitchen door.

Herbs

From a botanist's point of view, a true herb is a plant that does not produce a woody stem and that dies back to the ground each winter. But, generally speaking, gardeners have expanded the term *herbs* to include aromatic plants grown largely for their ability to flavor foods, for medicinal

A windowsill herb garden can flourish year-round in a sunny spot.

purposes, for cosmetic uses, and as ornamentals. They include woody and nonwoody annuals, perennials, biennials, bulbs, and grasses in their definition. The broad classification comprises hundreds of plants that you can grow in gardens or as part of landscaping.

Most herbs are easy to grow. They can be mixed in beds with other plants or have dedicated areas, but whichever approach you opt for, remember that they generally like a sunny location. The oils, which account for herbs' flavors, are produced in the greatest quantity when the plants receive six to eight hours of full sunlight each day.

Herbs will grow in any good garden soil, and they don't need a lot of extra fertilization. Highly fertile soils tend to produce excessive foliage that is poor in flavor. Most herbs like soil that is slightly acidic. Many herbs do very well as potted plants, growing on windowsills throughout the year. For potted herbs, watering from below works well because it provides adequate moisture without soaking the roots. After the outdoor growing season is over, enjoy dried herbs in fragrant potpourris and sachets, or grow them indoors in pots on sunny windowsills and use them for culinary purposes.

Trees

One thing I really miss, living at 9,200 feet above sea level, is the ability to have an orchard. On our Minnesota farm, we had about a dozen kinds of apples and several kinds of plums. With such an extensive variety, we were able to pick apples and plums for several months. What we didn't use, we shared with friends and neighbors—and what they didn't use, our livestock enjoyed immensely.

Trees (and shrubs) provide shade and wind protection; they yield not only fruits but also nuts, lumber, firewood, and fence posts. They invite wildlife to share our world. Planting them and nurturing them is really not all that hard, and the rewards are great.

Container-Grown versus Bare-Root Stock

Trees are usually sold either as container-grown, balled-and-burlapped (B&B), or bare-root stock. Container-grown trees are the easiest to plant and successfully establish in any season, including summer. With container-grown stock, the plant has been growing in a container for a period of time, so little damage will be done to the roots as the plant is transferred to the soil.

One disadvantage to container-grown trees is that they may be *root-bound* if the nursery kept them in the same size pots for too long, and root-bound plants may suffer from problems. When you remove root-bound

An Iowa farmer prepares to plant a row of B&B evergreens as a windbreak.

plants from their containers, you will notice that the roots are wrapped tightly in circles at the edge of the soil they were planted in. You can help these plants get a better start by pulling the bound roots apart and spreading them out in the planting hole.

Growers dig B&B trees from the soil, wrap the roots and soil in burlap, and keep them in the nursery for an additional period of time to give their roots an opportunity to regenerate. B&B plants can be quite large. It's best to transplant B&B plants in the fall or early spring while they are dormant; they do not do well with summer transplanting without a tremendous amount of babying.

Bare-root plants are usually extremely small. Because there is no soil on the roots, you must plant them when they are dormant to avoid their drying out. Seed and nursery mail-order catalogs and wholesale traders frequently offer bare-root trees. Many state-operated nurseries and local conservation districts also sell bare-root stock in bulk quantities for only a few cents per plant. Bare-root plants usually are offered in the early spring and should be planted as soon as possible upon delivery.

Why start small? That depends. It is tempting to start with a larger tree and shorten the time until first harvest or measurable shade. However, sometimes the smaller bare-root trees grow faster than larger ones whose foliage is too much for the disturbed roots to support. So, if you opt for larger specimens, many trees benefit from pruning at transplanting time to reduce the demand on the roots, which are in the process of establishing themselves. Again, consult someone who is familiar with fruit and ornamental tree husbandry to give your tree planting endeavor the greatest possibility of success.

Fruit and Nut Trees

In order to produce, most fruit and nut trees have a chilling requirement, which is defined as the accumulation of hours that the tree needs to spend below 45 degrees Fahrenheit and above 32

Apple trees with ripe fruit, ready for harvest.

degrees Fahrenheit during the course of the year. The requirement varies by species and variety. For example, apples require between 250 and 1,700 hours, depending on the variety, and blueberries require 250 to 1,200 hours. After an unusually mild winter, or an unusually severe winter, you may notice that fruit crops are poor despite the fact that the variety normally performs well in your hardiness zone. This is simply because that particular winter did not meet the tree's chill requirements.

Some fruit and nut trees, like apple and pear, need cross-pollination with another tree of the same species, although the tree must be of a different variety (in other words, a Jonah apple and a Jonah apple can't cross-pollinate, but a Jonah and a Red Delicious can). Nursery stock for some trees that need cross-pollination may come with two varieties grafted onto the same root, thereby allowing a single tree to cross-pollinate. Other trees, such as apricot and sour cherry, are self-pollinators, so male flowers pollinate female flowers on one tree by pollen from the same tree. In a few fruit

Young cashews on a tree.

species, such as kiwifruit and persimmon, different plants produce the male and female flowers. In these species, only the female plants bear fruit, although a male plant must be nearby to produce compatible pollen. Many fruit trees are available in standard, dwarf, and semidwarf sizes. The standard-size trees produce more fruit and live longer, but they also take longer to begin producing. They can reach heights of 30 or more feet and need to be spaced adequately so that they are not competing with other trees for moisture and sun. The taller a standard-size tree grows, the harder it is to reach the top branches for harvesting fruit, pruning, and pest control.

Dwarfs and semidwarfs are created by grafting cultivars onto root stock whose genetic potential for height is much lower than full-sized varieties—anywhere from 5 to 15 feet as a general rule. Dwarf and semi-dwarf varieties can be trained along a fence or wall on an *espalier*; this provides fruit in small areas, such as a backyard. One caveat about dwarf/semidwarf varieties: any suckers sprouting from the roots must be pruned off because they will not produce quality fruit. Also, make sure that the graft is above soil level when the tree is planted.

Most fruit and nut trees benefit from pruning. Without it, branches become crowded, thus reducing airflow and making the leaves and fruit prone to fungus and/or disease. Without proper pruning, fruit may be plentiful but undersized.

Different trees need different types of pruning. Some do better with a central leader—a strong central trunk from which limbs branch. Others, such as peach trees, do well with an open center, created by pruning out the central leader, which opens a tree's crown.

You should prune after the coldest part of the winter but before trees break dormancy. A general rule of thumb is to prune

Grapes and Cane Fruits

Propagate grapes and cane fruits fairly easily from cuttings off existing plants or by layering. Nursery stock is also readily available. Spring is the best time for planting as well as for pruning, which you should do before the plants break dormancy. Most cane and vine fruits bear the best fruit on new or one-year-old shoots, so pruning out older shoots will help production. Grapes, blackberries, blueberries, currants, gooseberries, and raspberries all grow on canes or vines, are typically long lived, and are spread via new runners from the root system. There are varieties that work in most plant hardiness zones.

between 20 and 50 percent of the previous year's growth. Eliminate any broken or crossed branches, and cut out the "water sprouts" (vigorous upright growth springing from the trunk and branches) because they will not bear fruit and will serve only to shade the tree and draw off nutrients.

Before spring begins to coax buds and blossoms into being, spray fruit trees with a dormant oil in order to smother all overwintering insects that may attack the tree once summer comes. The cruel truth about growing your own fruit is that you are not the only one who wants it. Bugs do. Caterpillars do. More creepy-crawly beasties than you want to know about can potentially invade your little corner of Eden. This is what makes organic growing challenging because bugs don't back down willingly. If you do nothing, don't be surprised if your apples turn out wormy, your peach trees succumb to borers, and fire blight reduces your pear trees to black skeletons. For beautiful homegrown organic produce, you'll have to do your research, take advantage of the plethora of information and products out there, and fight the good fight. It will be worth it when you hold in your hand the literal fruit of your labors.

Forage Crops

Forage crops are often the first field crops that new farmers need to deal with since livestock often plays a role in the life of small farmers. Like everything else in the world, forage has its own lingo.

Pasture. A field of grasses, legumes, and forbs (broad-leaf plants) that farmers grow to feed livestock. Although pasture may require some maintenance (liming, spreading manure for fertility, brush hogging to reduce weed and brush pressure), it is not routinely tilled.

Hayfields. Cropped fields that farmers periodically till, plant (with grasses and legumes), and harvest, producing bales of dry forage. You can easily put up small quantities of loose hay—say,

Preparing Haylage and Silage

Prepare both silage and haylage by packing the harvested crop into airtight containers (silos or large agricultural plastic bags in the field) so that bacteria ferment the feed. Although the upfront costs for an operation that allows haylage- and silage-making are stiff, these processes offer benefits for larger operations. Making your own haylage and silage is cost-effective when dealing with large amounts of feed, when weather and crop moisture are more forgiving for ensiling than for haymaking, and when you can chop and ensile stubby, drought-stricken crops that would make poor hay.

enough to feed a couple of sheep or some rabbits—rather than making bales. It must, however, be stored where it can't be rained or snowed on, or it will become moldy.

Straw. Hay and straw are not the same thing. Hay is made from grasses and legumes, cut before the plants have gone to seed and died back. Straw is the residual dead vegetation that a grain crop, such as wheat, rye, rice, or oats, leaves in a field after harvest. Farmers bale this dry material as they do hay, but, compared to hay, straw has few nutrients. Therefore, it is mostly used for bedding, mulching, or straw-bale construction rather than for feed.

Haylage. An *ensiled* feed (stored in a silo or other airtight closed container) for livestock, prepared from cuttings off the hay field. You can first bale it and then bag it, or you can chop and blow it directly into long, tubular bags. You can make haylage when crop moisture is between 40 and 60 percent.

A grain field after harvest.

Silage. Another ensiled feed, prepared from crops like corn or sorghum. Farmers ensile such crops because their stems are dense and retain moisture. You can put silage in a silo with moisture content between 60 to 70 percent.

Making Hay

Properly harvested and stored, hay can last for extended periods of time with the bulk of its nutrients intact. Hay that is baled with too much moisture will mold and, if stacked, may begin to heat up like a compost pile, sometimes to the point of spontaneous combustion. Hay should have less than 15–20 percent moisture when it is baled or piled. Hay stored out in the elements tends to mold as well as lose valuable nutrients. Horses, especially, cannot eat moldy hay. The quality of hay is dependent on several factors:

- The moisture content at baling and stacking
- The time in storage and storage conditions
- The age of the hay when cut and baled (did you catch it at its peak?)
- What cutting you are taking (first cutting of the season, second, and so on)
- The species present (not only the grass or legumes but also the weeds)

The optimal time to cut hay is when grasses are just beginning to reach the flowering stage but before they begin setting seed. This is the time when the nutritional value of the plants is at its peak. Hay cut with a sickle mower or by hand with a scythe is *windrowed* (loosely piled in long rows to dry before baling). Windrowing is usually done with a PTO-operated hay rake, although it can be done by hand for a small patch. Hay cut with a haybine or swather is dropped by the machine into windrows, thus reducing your workload. However, swather windrows tend to be more compacted and slower to dry than those rolled by tractor-drawn rakes.

Bales of hay provide forage for livestock.

The great constant inconstant that all farmers come to terms with sooner or later is the weather. We can't control it. We can only roll with it. As with all other crops, how well your hay turns out depends largely on the weather. If you have a few nice, sunny, breezy days after cutting, the hay dries quickly and the windrows don't have to be turned before baling. But if clouds roll in and even a little moisture falls, the process can take longer, and you may need to turn the rows with a rake several times before they are ready to bale. Each time you have to turn windrows, a certain amount of leaves and blooms breaks off, reducing the nutritional value. But, as previously noted, baling wet hay is worse than baling bleached or stemmy hay.

Pests

As I've mentioned, you won't be the only one who likes your plants: animals, insects, and microorganisms will find your vegetables, fruits, and flowers to their liking as well. Although commercial agriculture has come to depend on harsh chemicals to control pests, it's best to resist the temptation to use these chemicals. Synthetic controls reduce biodiversity, thereby killing off the good organisms that work on your behalf. In addition, they are often dangerous to you, your family, and your livestock. Whenever possible, go organic.

Many farmers are raising crops successfully without toxic chemicals. Your first line of defense is a healthy garden; well-cared-for plants are much less susceptible to pests than an overgrown garden, so keep up with weeding, pruning, and other chores. Catch problems early by performing regular checks. You'll catch the most pests by checking at different times of the day and night, and dawn and dusk are key times when many unwanted and unsavory characters are busy in the garden. As old Ben Franklin advised, "An ounce of prevention is worth a pound of cure." Try these approaches to animal, insect, and disease control:

- Use manual methods, such as regularly pulling and hoeing weeds or picking off and disposing of pests or infested leaves to keep the situation under control. Do not add diseased or infested plants to your compost pile lest you proliferate your problems.
- Practice crop rotations, even in the garden. Many pests become a problem only when you grow the same crop in the same place for years. For example, once tomato plants start succumbing to blight, each year will be progressively worse if you plant them in the same place because the spores are in the adjacent soil.
- You can attract pill bugs (also called sow bugs and roly-polys), which love ripe fruit, to cut-up pieces of raw potato placed on the ground and then drown the bugs in a bucket of soapy water. I know it sounds heartless, but it gets slightly easier when you witness the amount of strawberries, tomatoes, and other produce that these little roly-polys can damage. If it feels

less cruel, flush them down the toilet or let your chickens do the job for you. Insects are a great source of protein for them, and as long as they aren't at risk for predators, chickens are star members of the bug patrol.

- Fences can keep out large animals, such as deer, and electric wires strung low will deter rabbits and raccoons. Coons wear that little bandit mask for a reason! They can strip a newly ripened corn patch in a couple of nights, and they seem to know exactly when the ears are full and sweet.
- Your family dog can be a great deterrent for unwanted animals in the garden.
- Recognize that not all bugs are bad. Spiders, ladybugs, lacewings, and minute parasitoid wasps ("mini-wasps") are all bugs that eat bad insects in your garden. Create an environment that beneficial insects like, and they will control many problems for you. Plant asters, calendula, chervil, chrysanthemums, coriander, cosmos, dill, fennel, fleabane, poppies, Queen Anne's lace, rosemary, rudbeckia, sunflowers, sweet alyssum, and yarrow around the garden to attract these beneficial insects.
- Slugs and snails mostly crawl at night, when it's cooler. Keep them away from plants and produce by grinding up eggshells and sprinkling them in a ring around each plant. Powdered ginger sprinkled around a plant can also help keep slugs and snails at bay (remember to reapply after rain). Make traps by placing upside-down flowerpots, dark-colored plastic sheeting, and wooden boards around the garden. Collect snails and slugs in the early mornings from your traps and drown them in soapy water or visit your chicken coop with such offerings. Wrap a strip of copper around a tree trunk, a flowerpot, or the wooden sides of garden beds or fences. The unpleasant reaction between their bodies and the copper will repel snails and slugs.
- Diatomaceous earth sprinkled around the base of plants and on leaves will reduce populations of soft-bodied insects, such as aphids.
- Hose down plants and shrubs with water to rid them of aphids, whiteflies, and spider mites, being careful not to damage flowers or buds.
- Place a tin can with its ends cut off around plants to a depth of 1 inch to keep cutworms away or wrap stems of transplants with a strip of newspaper.
- Use agricultural fabrics to create a physical barrier that keeps insects away from plants. For example, floating row covers are very effective at protecting cruciferous vegetables from cabbage loopers. Nets can deter birds and deer from nipping your grapes, cherries, and the like.
- Brew your own bug sprays. For

Ladybugs are helpful insects in the garden, providing natural pest control.

A cabbage plant damaged by garden pests.

one such mixture, combine 100 mL of crushed hot pepper with 400 mL of water. Strain the concoction and spray it on infested plants. (Note that hot pepper can irritate your eyes and skin.) Do not spray when it is windy or in strong midday sun. Or finely chop ten to fifteen cloves of garlic and soak them in 500 mL of mineral oil for twenty-four hours. Strain the solution and spray as is, or dilute it with water before applying. Be aware that use of these solutions may also repel beneficial insects.

- To help control grubs, use commercially available nematodes, which are microscopic worms that prey on grubs and other pests.
- Place bat houses and birdhouses near your garden. It is nothing short of incredible how many insects they can eat! In some parts of the country, farmers set up multiple houses for purple martins, the largest member of the swallow family. These seasonal birds are not only friendly songsters and aerial acrobats, they eat only insects.
- Use companion planting because many pests have a natural aversion to mint, garlic, basil, chives, dill, onions, marigolds, and other aromatic plants. These plants may be interspersed with more vulnerable plants in your garden.
- For bad infestations, try using insecticidal soap (dissolve one part dishwashing liquid in forty to eighty parts water). Spray infected plants, covering the undersides of leaves, and rinse off after fifteen minutes to avoid damage to foliage. For severe infestations, repeat three times during a ten-day period to treat successive generations. As with anything you spray onto plants, use caution as temperatures climb; topical applications clog pores on the plants' surfaces, stressing them in hot temperatures.
- Bacillus thuringiensis (Bt) is a commercially available, naturally occurring soil bacterium that infects and kills caterpillars when they eat it. It's used for many garden pests, including cabbageworms, cutworms, tomato hornworms, tent caterpillars, and gypsy moths. You'll find Bt in garden centers as "organic garden spray." According to the Colorado State University

Extension, Bt is unusual in that it does not have a broad spectrum of activity, thus beneficial insects are safe. In addition, Bt is essentially nontoxic to people, wildlife, and pets. However, Bt-based products tend to have a shorter shelf life than other insecticides.

Adding Flower Beds

In my opinion, it's hard to have too many flower beds. There are so many kinds of flowers and so much beauty to create. I love the particular joy found in transforming an overgrown corner of the yard into an outdoor oasis or bringing color, purpose, and magic to what was a nondescript bit of property. In earlier years, I used more of a fly-by-the-seat-of-my-pants approach to gardening, but trial and error is a tiring way to go! Through the years, I have learned to see the value in planning. Here, then, are some flower-bed basics so that if you aren't already an expert, you can cut down on the trial-and-error part of the process.

Master Plan

Although it can sometimes feel slow and frustrating, designing a flower bed on paper can produce superior results, and don't we all want success for our labor? You can just buy a bunch of flowers and plant them, and you'll have color, at least for that growing season, but creating an overview of the process can put you firmly in the driver's seat. So, first things first: ask yourself why. Why a flower bed? Why here? What do I want out of this particular spot?

The second step in planning is figuring out the cost in both money and time according to your wants and needs. For example, if you need the garden to be fully flourishing in time for your daughter's wedding, you might spend more up-front on plants and various elements than if your goal is to create something of beauty in your own sweet time. Proposing a reasonable budget can aid you in making decisions and prevent you from plunking down more money than you really want to. For example, if you decide that you can spend no more than $200 the first year, it will help you determine which variables you absolutely must have and which might be able to wait for another year or an unexpected check in the mail. A three- or even five-year plan that outlines what you can afford now and what you will add in the years to come is also a good idea.

Next, think about maintenance. Designing your garden is exhilarating, but when

How much effort do you want to put into maintaining your flower garden?

the weeds are in a pitched battle with your flowers, the borders and edges are getting ragged, and the temperature hovers at 100 degrees Fahrenheit with 90 percent humidity, you're going to be really glad that you didn't skip this stage. Think long term.

If you're new to gardening, you may not realize how much time certain tasks take. There's no such thing as a "no-maintenance" garden unless it's made of concrete and sports silk flowers. Take your level of experience into consideration. If you plan an extravagant flower bed and have never done much gardening, you may find yourself spending hours each week keeping it looking nice—hours you don't really want to spare. A good rule of thumb to help figure out what you can maintain is to estimate how much time it takes you to tend a certain square footage of flower garden per week at your present skill level. Let's figure that it can take an experienced gardener about an hour to maintain a 300–400-square-foot established garden at a fairly high standard of neatness. An hour a week is not much time—fifteen minutes here and there. So you could theoretically add another flower bed of similar size and spend around half an hour a day keeping both in great shape and even less time if you have someone helping you.

Assessing Your Site

It's time to bring your gardening goals to the forefront because they will help you finalize the perfect spot. If you want a place to de-stress, which spot is already most conducive to that? In her excellent book, *Designing Your Gardens and Landscape,* Janet Macunovich suggests placing your flower bed in a place you're most likely to see when you are stressed. Coming home from work? From your kitchen window? If you've put fragrance and butterfly/bird watching on top of your list, you'll want your flower bed to be situated where you can smell and see it.

Take the time to create a site assessment sheet, listing first of all what you can see from the main vantage point. Next, note the overall background as well as each existing feature (i.e., tree, barn, fence), including its shape, color, texture, and overall effect. Take note of the existing strong lines (sometimes referred to as "hardscaping") along with the overall feeling you get from the site. Mapping out what is there can help you decide what to keep, what to eliminate, and how to camouflage aspects that you can't change.

Consider your location for both aesthetic value and ease of care.

Now it's time to make a sketch with either a pencil and paper or a computer design program (some are made specifically for garden design). A ground-level view is helpful at this stage. An eagle's eye view—straight down on the layout, similar to a building's blueprints (a

Testing for pH

"Must have acidic soil." "Prefers more alkaline soil." We see these types of statements on plant labels, so here is a primer for pH. The University of Maryland (UM) Extension has this to say: "It [pH] plays a big role in the availability of nutrients to plant roots, nutrient runoff and leaching, and microbial efficiency."

The pH scale operates like the Richter scale for earthquakes. The United States Environmental Protection Agency (EPA) says, "A pH of 7 is neutral. A pH less than 7 is acidic, and a pH greater than 7 is basic. Each whole pH value below 7 is ten times more acidic than the next higher value. For example, a pH of 4 is ten times more acidic than a pH of 5 and 100 times (10 times 10) more acidic than a pH of 6. The same holds true for pH values above 7, each of which is ten times more alkaline—another way to say basic—than the next lower whole value. For example, a pH of 10 is ten times more alkaline than a pH of 9."

Pure water has a neutral pH of 7.0. You may be able to lower your soil's pH by a decimal point or two over a period of years by raising the flower bed above the surrounding gardens/yard and by adding organic matter. To move it toward base (alkaline), add agricultural lime (calcium carbonate). Soils generally reflect the pH of their surrounding geographic areas, so looking at native plants will give you a ballpark idea of the soil pH in your yard; for example, a dominance of sweet clover can indicate a soil tending toward alkaline, whereas thriving blueberries point toward acidic soil.

plan view)—can help you map the location of existing structures and give you a base from which to begin designing. Cornell University Cooperative Extension states, "Knowing the size of your garden or landscape area has value in making future decisions. Determining the square footage is necessary for deciding on materials to purchase, including grass seed, fertilizer, compost, mulch, [and] number of plants to cover an area." Match the design layout with the realities of the site size.

Another critical aspect, according to the Cornell Extension, is to investigate obstructions below ground level. Where are your water, gas, and electric lines? As someone who just last week hacked a large hole in the PVC water main with my enthusiastic wielding of a pickax in pursuit of some stubborn bush roots, take it from me—just because you can't see these things doesn't mean that they don't exist! Who knew that the line connecting the sprinkler control box to the water main was literally only 6 inches below the surface? Certainly not me. Let's just say that my afternoon took a completely different turn, and I will no longer do any digging until I first make a map of the area.

While choosing your site, assess such things as sunlight, soil type, water, exposure, and competition from roots.

Sun

The University of Illinois Extension advises that the best way to assess how much sun your prospective site actually receives is to observe the sun pattern. You may be surprised at how much or how little sunshine any given area gets. Because each plant has certain requirements, the success of your new flower bed lies in knowing which plants work in which type of area.

- **Full sun.** Most sources agree that six or more hours of direct sunlight in a 24-hour period qualifies as "full sun." The six hours don't have to be continuous; there may be midday shady periods. As long as your site receives at least six hours of direct, full sunshine, it fits this category.
- **Partial sun.** Between four and six hours of direct sun.

- **Partial shade.** Less direct sun than partial sun, but more partial sun than shade: approximately two to four hours per day.
- **Shade.** Less than two hours of direct sunlight per day.

Soil

While soil is an amazing substance, remember that it is not always in its optimal state. Yours may be sandy, loamy, silty, or full of clay. Soil can be amended, but you need to know where you are starting. Don't just assume that topsoil you might bring in is an improvement over what you already have—the texture may be better but the nutrient levels lower, for example. So test what you've got so you know what you need in terms of soil amendments.

Note also if the ground is loose, firm, or compacted and in need of aeration. Clay soil can be higher in nutrients and retain water better than sandy soils. The drawback for those of us who have too much clay is the shortage of air to the plants' roots. If there are plants already growing there, are they in good health? If not, why? Are their roots shallow or deep?

Water

It's a basic fact that plants need water. However, there are some specific aspects that you'll want to include on your site assessment. First, how much natural water does the site get? Are some areas blocked by overhanging eaves, trees, or buildings? How high is your water table (if it's quite high, moisture will wick upward and benefit your garden)? Next, consider how you will irrigate your garden: will you use jets, misters, soakers? Will you operate your irrigation manually or program it to come on automatically at certain times? Also consider how much runoff your site gets and how serious of a problem it is. Finally, does your site collect standing water? Some plants can handle standing water, but if it's a chronic problem, you may need to raise the bed or correct it by some other means.

If you plan to water your flower beds by hand, make sure that you can easily get the water to the flowers.

Exposure

Consider your prospective site's exposure to weather and disturbances. Are you in a frost zone? Is your site in a windy area? How much heat does your site get? Certain sides of the house will be significantly hotter than others at certain times of the day. What about children, animals, foot traffic? The shortest way between two points is a straight line, and kids will use that approach even if it means cutting right through your flower bed.

Root Competition

Keep in mind that trees not only lend their shade and beauty, they also spread out their roots usually at least as wide as their canopy. Those tree roots, as well as any roots from bushes and other existing plants, will be competing with your flowers for water and nutrients.

Choosing Plants

By now, you've created a site assessment sheet that includes your goals for the proposed flower bed, projected budget, drainage quality, and soil pH. It's time to select which plants will do best in your spot. At this point, you'll decide whether to go with a complex or simple layout based on your preferences. A complex garden—one that must meet many different criteria—is obviously a bit more challenging.

Macunovich advises gardeners to start with ten plant choices. You can build a whole bed with multiple plantings of those ten plants, and too many plants at once can be overwhelming. If you decide you want more than ten different plants in your flower bed, look for more after you've settled on the first ten. Record your reasons for selecting each plant; for example, "lavender: fragrant perennial, heat and drought tolerant, cut flowers, 1–2½ feet tall, gray-green foliage, purple blooms." And, while you're selecting plants, don't limit yourself to flowers alone. Foliage is incredibly diverse and can echo or complement favorite flower varieties or furnish a stunning focal point all on its own.

One dilemma that gardeners encounter is whether to go with annuals or perennials. Annuals typically have a long blooming season and cost less than perennials of equal size. However, perennials recoup those costs because most live for at least three years, if not longer. The main drawback for perennials is that they bloom for shorter times, and they may bloom earlier or later than most annuals. If you go with a perennial garden, you have to consider how to keep your garden looking good and growing throughout the season by overlapping blooming times so that you can maintain a constant display of color. Because this can take a while to establish, at least at first, you might opt for a mixture of annual and perennial plants if doing so aligns with your goals for that particular flower bed.

If your goals include a specific color or color scheme, that will help point you toward certain plants and away from others. If you're on the crest of the decorate-your-outdoor-space wave, you might select plant colors that echo or draw out the furnishings for your patio area (or vice versa). A garden seen

Color of the flowers is one of the top considerations when selecting plants.

Edgings

Edgings can act as barriers against root invasion, give definition to your flower bed, create continuity by using materials that are seen elsewhere on your property, and/or help draw the viewer's eye toward the focal point(s). This can be as simple as an "English border" edge, which is a small ditch between lawn and flower bed with the lawn side cut straight down and soil sloped up onto the flowerbed side. This straight cut allows you to see when weeds or roots start to reach their greedy little tendrils toward your pristine flower bed, and you can nip them off before they get too invasive.

Beyond such natural edging, materials abound—bricks, rocks, wood, cement. For weed control, the edging should be buried deep enough to come between those opportunistic and unwanted roots and your flower bed. When deciding which material to use, aim for continuity. For example, if there are no bricks anywhere else on your property, they may look out of place as an edging, especially if your flower bed is informal.

mostly in dim light will be more dramatic with whites, yellows, and pastels to reflect light. If you're aiming for curb appeal, echoing your trim color can be subtle and effective, made more dramatic with a contrasting color.

Diversity in foliage—shape, color, texture—can carry you through. Another technique, sometimes referred to as *sibling planting*, uses the subtle similarities between plants—those of texture, leaf shape, bloom shape, color, and size—to give continuity to the garden without a repetitive feel. For example, the low, mounded whiteness of sweet alyssum along a border may anchor a bank of tall white daisies or white Echinacea glowing in a back corner. The deep red of a Japanese maple echoed in bronzed-leaved, burgundy-flowered dahlias splashed against the ground captivates our gaze and keeps us searching for more such serendipitous pairings. And while you're making plant selections, don't neglect ornamental grasses, which come is all sizes, colors, and habits.

Bringing it All Together

Initial planning helped you select your main viewpoint. Now's the time to establish the focal point, which is a spot within the garden that will draw attention, provide balance, and help the viewer align with an attractive feature. Choosing a focal point at the very center of your site will give your garden a more formal look, lending itself to stricter symmetry. In contrast, a naturally wooded backyard or one with trees and bushes may lend itself to the informal look of an off-center focal point. To which background feature do you want viewers' eyes to be drawn? Your front door? A water feature? Place your focal point between you and that important background.

Watch for negative background elements as well. For example, if your neighbor has parked an old car on his side of your property line, and you're adding a flower bed between your house and said vehicle, you'll draw less attention to the slowly rusting relic by directing focus to the right or left of a direct line of sight. This way, viewers may not even notice the car.

Drawing it Out

Now you're ready to make a scale drawing, either by hand (you do *not* have to be an artist) or via software. First, decide on the scale. You should have the measurements from your original site map. It can be helpful to establish a landmark from which to measure. This can be the edge of your patio or sidewalk, for example, because straight lines are easiest to utilize. Without this frame of reference, a new flower bed could end up situated at any kind of angle at any point in your yard. Create a grid on your paper, taking measurements from the landmark to find your lines of orientation.

Next, pinpoint the focal point in its actual location in your prospective flower bed, then go outside to look at it from your viewpoint. If you have a helper, that person can stand in the position of the focal point. If you're minus a helper, utilize some other object, such as a shovel or bucket, and position it where you want your focal point to be. Once you're satisfied with its placement, measure the distance to the right or left between your focal point and your landmark and the distance above or below the landmark from the focal point, and transfer these numbers onto your plan. Drawing an arrow from the viewpoint pointing toward the focal point can help keep you oriented as you begin to add other elements to your plan.

Here comes one of my favorite parts: which plant, out of all that you've got on your "wanted" list, is best for this focal point I keep mentioning? You might choose the most unusual or colorful plant, or maybe the tallest plant, but the placement itself is what makes it focal. Consider longer blooming plants or plants with foliage, texture, or shape to carry them through nonflowering periods. If your focal point is large enough, a combination of plants can provide just the right look.

Macunovich recommends that the plant or grouping in your focal point be at least as wide as 5 percent of the distance between the main viewer and the focal point. For example, if the main viewer will be sitting 20 feet away from the focal point, the plant should be at least 1 foot wide. At 50 feet away, you'll need a plant or grouping of at least 2½ feet wide. A patch of very tall plants can function as well as a wider group of short plants, depending on the setting.

Macunovich further suggests using objects of varying sizes to ascertain how large a plant must be to have the desired impact (and not lose itself among all the other plants) and how much space each

of your chosen varieties will occupy. If you're using pencil and paper rather than a computer program, cutting out circles of varying sizes representing those plants will give you an easy way to manipulate the shapes until you find the positioning you like. Computer programs make this stage very easy as well. Then go ahead and draw them in and make note of how many of each type you will be purchasing. Since plants often don't reach full potential until

Your garden can be a place of serenity where you can relax and enjoy the fragrant blooms.

Rock formations and interesting shapes add visual appeal.

the end of their season, choosing how many plants to put in the ground is where it can be tempting to go beyond your budget rather than waiting for your selections to grow and fill in the blank spaces.

To help offset that impulse, keep in mind that empty space is not necessarily bad. Look at your design plan as you would a picture you're painting, visualizing where it might need filling in and where space actually enhances the effect you're aiming for. If, from the main viewpoint, the overall garden feels finished, you can decide to let the plants fill in the space as they grow. If, however, you get an unfinished feel, perhaps you will want to fill in a few of the bare spots.

Also keep in mind that you will need some way to access the inner plants for care and maintenance (unless the bed itself is small enough that you can reach the middle from the edges). Leaving a bit of space between groupings can make maintenance much easier.

Choose plants to frame your focal point, which is one way to make it even more focal. So choose a plant (or plants) that is significantly different in some aspect, such as flower or foliage color, shape, height, or texture. This will cause your focal point to "pop," similar to the effect that a mat and frame has on a picture. You can combine plants in an informal or formal arrangement to make that frame, leading the viewers' eyes in such a way as to create a visual, if not literal, path to that spot.

No edge may be needed if your bed fills a spot between a sidewalk and house or driveway. But for other locations, the outer shape of the bed is a subtle yet important aspect of its overall design. That said, there is no right or wrong outline. Just as with the other elements, whatever you do needs to draw the viewer's eye toward, rather than away from, your focal point(s). For curved outlines, here are a couple of general guidelines as you explore first on paper and then on the actual site.

- "Dips" in the outline should not be deeper than they are wide (except on very large beds with curves of 15 or 20 feet).
- Inward curves tend to draw attention inward, so consider using them to draw attention toward important features.
- Check the outline from multiple vantage points. What works at viewpoint A may not look so great from viewpoint B or C.

Final Adjustments

To make sure that you're pleased with your design, take a mental tour of your proposed site while imagining the plants as they appear on your plan. How will your important plants and groupings look from each vantage point? Are their backgrounds appropriate? Do lines lead the viewer's eye only to attractive features? Is each view balanced? Is this garden set up in a way that it will be easy and enjoyable to maintain? And, finally, does it meet your initial goals?

Changing things at this stage is as easy as using an eraser or deleting a shape on the computer screen. Making a line around each grouping of flowers/plants can be a helpful way to assess the effectiveness of your design arrangement because those lines draw your gaze to their endpoints. Use your eraser and pencil or software to shift plants to break up or redirect any less-than-pleasing lines, if needed. Mentally weighing the garden as a whole can feel a bit challenging, but you need to do a final assessment and make any adjustments now because once you start planting, it will be more difficult to make changes.

Saving Seed

The Why: Saving seed from open-cultivated heirloom plants goes hand in hand with growing them. Not only does it make economic sense, it ensures that you will always have seed from your favorite cultivars, even if seed catalogs stop carrying them. It also connects you more strongly to the whole cycle of growing, harvesting, and growing again.

The How-To: Some plants are self-pollinating, but even those can cross-pollinate via insects or wind. Since the goal in growing heirloom varieties is to preserve them in their original form, prevent them from crossing with some other variety. If your garden is small, consider growing only one variety of each species so that there is no danger of cross-pollination. Another, although not foolproof, option is to isolate cultivars by separating them by 10 or more feet. Yet other methods are to grow them in screened cages or to cover individual flowers with bags and hand-pollinate them (laborious, yes, but possible). *Time isolation*, which means to time planting so that different varieties don't flower at the same time and therefore can't pollinate each other, also works. Here are the basic steps to follow to save seed:

1. Choose your healthiest, most productive, and most flavorful plants to save for seed and harvest the rest. Prime specimens produce healthier seeds than their weaker, stressed counterparts.

2. To help ensure that your seed will grow well, allow chosen veggies/fruit to ripen fully. Depending on the vegetable or fruit you're growing, you may leave them on the plant until dry and wrinkled. Seeds harvested before they are fully ripe do not grow well.

3. Provide warm, dry conditions while the seed matures to help increase storage life. Especially if rain threatens, many seed savers will harvest the produce if fully ripe and then bring it inside for final drying.

4. Separate seeds from the flesh (this is easier for some plants than others).

5. Make sure that your seed is completely dry because damp seed may mold. Place the dry seed in a tightly closed jar and store in a cool and dry location. Refrigeration will increase life expectancy. A bit of diatomaceous earth in the jar will keep insects from damaging your precious seed, and a packet of silica gel can help keep it dry.

6. Label your jar clearly to avoid confusion next spring.

7. When processed and stored properly, most vegetable seeds remain viable for three to five years. You can test for germination by sprouting a few seeds between moist paper towels and the side of a glass jar, just like in your third-grade science experiment. If germination is low, plant extra to ensure that you'll have the amount of plants you want to grow.

Trends in Planting

Some things never change. True. But then again, some things do, including what's trending in the garden and yard. Whether for personal pleasure or entrepreneurial ventures, hobby farmers and gardeners across the nation are exploring new, repurposed, or rediscovered favorites in their own backyards.

Heirloom Varieties

Heirlooms are not just the furniture pieces you inherited from your Great Auntie Pat. They're vegetables and fruits, too. In the world of gardening, heirlooms can be defined in different ways. One definition classifies heirlooms as cultivars that have been grown for a certain length of time (usually at least forty years). Some gardeners adhere to the definition that the only true heirlooms are grown from seed that, like the aforementioned furniture, have been passed down by a family or group who have preserved them. Or both. Whichever definition you adhere to, one thing is clear—heirlooms are not hybrids. A heirloom seed, when planted, will grow a plant exactly like the plant from which it originated.

People grow heirloom varietals for several reasons, one being that they are a taste of the past. Just like endangered species, some fruits and vegetables have been lost over time as fewer and fewer people save seed every year. Growing these old cultivars is a way to stay connected to our heritage, whether it's the Hopi Grey winter squash, almost lost to cultivation before its comeback, or the Anasazi bean found in a cliff dwelling and believed to be the same variety grown by the Hisotsonim cliff dwellers of the Colorado Plateau more than 800 years ago. Another bean, this one named the Mostoller Wild Goose bean and found in the craw of a wild goose, dates back to 1864 in Somerset County, Pennsylvania.

Many heirloom varieties spring from the efforts of gardeners who save seeds only from plants offering superior flavor and texture/tenderness. In a world where uniform appearance, ease of shipping, and adaptability to multiple growing zones have become goals, these objectives rob produce of taste and tenderness as growers stray far from original varieties into hybridism. Heirloom growers do just the opposite. As Clemson University Extension in South Carolina puts it, "When

Three varieties of heirloom tomato.

gardeners save the seed of the best tasting, best performing plants in their gardens each year for a number of years, they gradually select their own special cultivars. Furthermore, if grown in one locality for years, heirlooms adapt to climate and soil and can sometimes even out-produce hybrids. Those selections will be suited to their own growing conditions and tastes. Open-pollinated seed that has been grown and harvested for generations in a region or microclimate becomes adapted to that area's soil, climate, and pests."

This leads us to yet another reason that heirloom varieties are more than just nostalgia. Mankind needs their genetic traits. If old varietals are allowed to die out, the gene pool shrinks, and this can lead to increased disease and pest problems. We don't have to look very far to see how shortsightedness and hunger for the highest dollar often drive the planting of hybrid seed. This, in turn, has led to the extensive use of pesticides and other chemicals to help cope with the disease susceptibility of widely grown cultivars that are selected not for taste, texture, or resistance but for productivity, appearance, and how well they will ship. Mankind can't afford to let the plant gene pool diminish. Who knows if some obscure heirloom variety may hold the genetics for life-saving medicine or to rescue a whole race from starvation?

Another reason to explore the world of heirloom cultivars is quite simply that it saves you money. If you save seed (see sidebar on page 123) , you avoid shelling out for new—and often expensive— hybrid seed. Since hybrids won't produce the plant they grew from, gardeners who go the hybrid route lock themselves into needing to purchase seed year after year, which is not cheap.

To be fair to the food industry, heirloom varieties do present problems for mass cultivation. For one, they don't mature all at the same time, which makes mechanized harvest next to impossible. However, this attribute actually suits smaller farmers just fine, giving them an extended harvest for home use. Rather than a bumper crop ripening all at once, heirloom varieties lend themselves to gathering just enough for dinner or for a canner full of whatever is ripe. Heirloom types often don't ship well, and they may look different from what consumers are used to seeing. However, shoppers who frequent farmers' markets are a savvy bunch, increasingly familiar with and desirous of heirloom produce, and they'll be looking for these old favorites.

Seed Life

As we can see by the Hopi squash or the Anasazi bean, some seeds last longer than others. Those that remain viable after five or more years are considered long-lived. According to Clemson University, this category includes "beets; all cabbage relatives such as broccoli, cauliflower, collards, and kale; cucumber; lettuce; melons; peppers; sunflower; tomato; and turnip." Under medium-lived seeds—lasting at least three years—Clemson lists beans, carrot, chard, eggplant, parsley, peas, pumpkin, and squash. Short-lived seeds, which include corn, leek, onion, and spinach, can be depended on only to last to the next growing season, although freezing seeds can extend their life somewhat if no moisture reaches them.

Distinctively dappled Anasazi beans.

Home Food Gardens

Hand in glove with heirloom cultivation, we are also experiencing a resurgence of home vegetable gardens. Evocative of the "victory gardens" of World War II, and in fact sometimes referred to by that name, people are planting food instead of grass or flowers in their little bits of Eden. "Grow what you eat; eat what you grow" is the mantra, as each gardener, in his or her own way in his or her own corner of the world, pursues a sustainable permaculture with strong overtones of biodiverse thinking. Often part of the modern homesteading movement, both urban and rural, home food gardens hark back to earlier eras as people take charge of what their family is eating from the roots up.

Although these victory gardens may not be linked with helping our nation win a war, as the first ones were, there is a growing revolution that has people from city to country taking personal responsibility and control for their food supplies once again. Ranging from growing tomatoes in patio containers all the way to classic gardens, people are rediscovering or reviving the wonder of being able to grow their own food in a sustainable way. Individual reasons to do so vary, but here are some that are driving this generation of victory gardeners:

- **Desire to be independent from corporate food systems.** Depending on a company, grocery store, or unknown and unauthenticated source for the food we eat puts us in a powerless position.
- **Concern about the overall health of foods we buy.** GMOs, pesticides, herbicides, foreign diseases, and insects riding in on imported food are just plain scary to contemplate. We care about the quality and safety of our food.

A garden of salad greens, reminiscent of the "victory gardens" of days past.

Hybrids versus GMOs

Hybrids and genetically modified organisms (GMOs) are not the same thing at all. Hybridization takes place when plants of the same species cross-pollinate. Tomatoes with tomatoes. Beans with beans. Corn with corn. But with the advent of genetic modification, scientists are splicing unlike species—tomatoes and salmon, soybeans and Brazil nuts, corn with genetically induced pesticides. Now that scientists are "playing God," we've got foreign combinations doing weird things in our digestive systems (i.e., altered gluten proteins that cause widespread gluten intolerance). The pesticide *Bacillus thuringiensis* (Bt), which paralyzes the digestive systems of insects, is married to the DNA of so many products that it's beginning to show up in human bloodstreams and gut tissues, contrary to its developers' prior claims that it would not do so. To make things even more complicated, GMO varieties can and will pollinate contributing species via air or insect-borne pollen so that even if one farmer doesn't plant GMOs, cross-pollination may contaminate his field; if so, the resulting crop will carry the GMO anomaly, further diluting and contaminating the gene pool.

- **Reducing the carbon footprint.** The average produce item travels approximately 1,500 miles before it arrives on your table! Between transport, refrigeration, pesticides, and fertilizers, that's a lot of petroleum.
- **Desire for a healthy and fulfilling lifestyle.** There's just something about being out in nature, working with the soil, breathing the air, absorbing the sun that it can sometimes feel like you're raising a crop of satisfaction and pleasure along with your tomatoes and peas.
- **Saving money.** Although in some areas, water bills can be an expense, if buying high-quality organic food is putting a strain on your pocketbook, growing your own can be a viable option to reducing your food budget.
- **Connecting to the source.** In a generation far removed from our agrarian roots, true sons and daughters are rising to take their places as knowledgeable, responsible, caring stewards of Planet Earth.
- **Return to more community-based living.** When people exchange their garden abundance with friends and neighbors, they sow much more than vegetables and fruits; in fact, they reap relationships on a level they may not have experienced before.

Conclusion? While grass may still be the most widely cultivated plant in most communities, watch out, lawns! Vegetables and fruits are coming.

Organic Gardening

Right along with growing heirloom varieties and victory gardens is growing organic. If you're raising your own produce to ensure the healthiness of what you and your family are eating, it only makes sense to go the final step and grow organic. The Environmental Protection Agency (EPA) gives this definition: "'Organically grown' food is food grown and processed using no synthetic fertilizers or pesticides. Pesticides derived from natural sources (e.g., biological pesticides) may [also] be used in producing organically grown food." Working with nature rather than against it, organic growing uses a variety of techniques to achieve good crop results without harming people or the environment. Organic methods include building soil structure and fertility and controlling weeds, pests, and diseases.

As you probably remember from earth science class, soil is the launching pad for what is grown;

it is the soil that infuses veggies and fruit with vitamins and minerals. Poor soil will grow inferior produce. Unfortunately, we know that some soils are a far cry from optimal—trampled down and inundated with years of chemical fertilizers, pesticides, and weed control. University of Georgia (UG) Extension says, "Organic gardening requires a long-term outlook with respect to soil preparation. In fact, the key to successful organic gardening is to feed the soil with organic matter, which feeds the plant, rather than to feed the plant with inorganic fertilizer as in conventional production."

Organic peppers at a farmer's market.

Green (plant) manure, legumes (nitrogen-fixing crops), and animal manures are further options to enrich your soil. Using bedding from your chicken coop or composted cattle or horse manure is a time-honored method. Just remember that animal manures in general and chicken manure in particular are very "hot"—meaning high in nitrogen—and can damage or even kill plants if applied directly. Composting the manure first is recommended, but if you don't compost, you can work around the nitrogen content by spreading chicken manure in the fall, after you've harvested your crops.

Keep in mind, though, that the manure itself must be free of pesticides and medications to actually be a positive addition to your soil. One hobby farmer who lived near a large poultry operation hauled several loads of manure and bedding from the grower and spread it on his garden. It was free. It was plentiful. But a couple of years later, when this same hobby farmer began to look into and apply organic principles to his farm, realization dawned that the chicken manure and bedding he'd been so diligent in spreading had been laced with the chemicals that the chickens had ingested. Once chemicals get into the soil, it takes years to remedy.

Growing a cover crop that you till into the soil increases organic matter, adding nutrients to the soil and decreasing the risk of certain diseases while helping to prevent weed growth in dormant areas. When used as green manure, small grains and grasses have been found to decrease the incidence of nematodes, microscopic worms that weaken certain plants by feeding on their roots. Legume crops "pull" nitrogen out of the air and fix it to their roots. The UG Extension states that you can add from 30 to 125 pounds of nitrogen per acre simply by tilling under your green-manure legume crop.

Timely cultivation and crop rotation contribute to success as well, and mulching helps reduce weed growth, conserve soil moisture and nutrients, and regulate soil temperature. Mulch also prevents erosion by limiting the amount of water that splashes on plants, which further helps reduce the spread of disease and keeps your produce cleaner; as the mulch decomposes, it adds organic matter to your soil.

Gardening without chemicals is an adventure. It's not just about planting and tending chemical-free produce. It's also a battle against the enemies of those plants—and, I will not lie, some pests and weeds are difficult to control. However, organic

To market produce as organic, growers must ascribe to chemical-free growing practices.

gardeners have a few tricks up their sleeves. For example, you can overcome certain soil-borne pathogens (bacteria, fungi, nematodes, and various viruses) and difficult-to-control weeds with soil solarization. Especially effective in warmer areas, soil solarization involves heating the soil to a sufficient temperature to kill the pathogens and weed seeds and is achieved by covering the ground with clear plastic for at least eight to twelve weeks. More temperate areas may require an entire year to rid the soil of these problems.

Why not use black plastic? As it turns out, clear plastic transfers light energy directly to the soil. Black plastic does absorb heat, but it also shades the soil, therefore preventing the temperature from rising as high as it needs to, thus taking much longer to solarize.

Organic pest control also uses natural products to kill or deter insects, fungi, and plant diseases. These products can range from diatomaceous earth and beneficial bacteria to insecticidal soaps and horticultural oils. Here some of the other ways that organic gardeners are fighting back:

- Planting resistant crops
- Timing planting carefully to miss insect cycles
- Rotating crops
- Practicing good cultivation, including responsible disposal of infected or infested growth
- Promoting pest predators and natural or biological means of pest control
- Increasing genetic diversity in cultivars

Fortunately for our planet, a whole lot of people are forging forward in the organic gardening world.

Superfoods

It's not surprising, with heirloom varieties and organic gardening on the rise, that a fresh generation of foodies is getting into the act, growing everything from quinoa to goji berries in their backyards or back pastures. We may not be able to grow acai (ah-SIGH-ee) berries, seeing as they come from the top of a type of tropical palm tree, but we can sure enough grow other "superfoods" (so called for qualities that enhance such things as brain function, heart health, good cholesterol levels, and more). Among these are berries—blueberries, elderberries, black currants, cherries. These familiar fruits pack an antioxidant punch, as do chokeberries—although, as Sandra Mason, extension educator at the University of Illinois, points out, "[Black] chokeberry, *Aronia melanocarpa,* needs a new name if it is ever to become a superfood!" This native North American fruit is being researched, and its antioxidant power may rival that of the aforementioned acai berry.

When it comes to superfoods, you'll find broccoli, spinach, and almost any kind of green on the list, and one of these—kale—is rated as one of the top veggies for absorbing free radicals in the body. Microgreens—spinach, kale, chard, and other greens harvested within fourteen days of their growth cycle—have been shown to contain four to six times as many beneficial nutrients as when full grown. Some microgreens can be harvested within five to ten days—talk about a short growing season! They're super easy to cultivate, too, lending themselves to container and year-round gardening.

And then there are sweet potatoes, which are high in fiber, antioxidants, and anti-inflammatory properties. They're not just for Thanksgiving anymore! For that matter, neither are pumpkins. Add them to the list of superfoods that you can grow in your own garden. Pumpkins are one of the richest sources of carotenoids, and the seeds as well as the flesh are good for you. While you're at it, you might as well plant a few beets, too. Although some people think that they taste like soil, beets have been called "nature's multivitamin." Not just their roots but also their leaves are edible and healthy, whether steamed, stewed, or juiced. And then there's garlic. This culinary staple is surprisingly easy and satisfying to grow, and it deserves the superfood label due to its natural antibiotic, antifungal, antimicrobial and detoxifying properties, as well as its cardiovascular benefits. Even the lowly cabbage has made the superfood list. Loaded with compounds that help detoxify our bodies of cancer-causing substances, it has been found to reduce cholesterol while providing anti-inflammatory nutrients and high levels of vitamins B and C.

Blueberries are one of the colorful superfoods, rich in antioxidants.

Among the lesser known superfoods that you can grow in your garden are sunchokes, also called Jerusalem artichokes. The tubers of this member of the sunflower genus, *Helianthus* (thus the "sun" in sunchoke), are high in heart-healthy minerals, vitamin B, and inulin, which acts as a prebiotic. Crunchier and sweeter than potatoes, they have a slight artichoke taste and are hardy in zones 2–10. Foodies are also growing goji berries, which are loaded with vitamins C and E, beta-carotene, and lycopene. This Chinese fruit is quite tart if eaten fresh, so it is usually dried like a raisin to give it a strong sweet-sour flavor profile. Goji shrubs grow well in zones 5–9 and can be trained onto a trellis or fence like other berries.

While you're growing exotics, how about growing your own chia seeds? Yup. Regular people are growing this pre-Columbian food staple in their home gardens. This is an annual plant, hardy to USDA zones 9–11, and you can take it through to the seed-saving stage so that you have your own perpetual source (although, realistically, if you eat a lot of them, that's a whole lot of seed saving), or you can merely sprout/soak purchased chia seeds in your kitchens to activate extra nutrition before mixing them into everything from smoothies to kombucha. Chia seeds contain omega-3 fatty acids that lower cholesterol and blood sugar while boosting energy.

Miniature Vegetables and Fruits

Mini veggies are on the rise in gardens both rural and urban. Sometimes called baby, midget, or dwarf, these small-sized garden goodies make tasty family and kid-friendly food or farmers' market offerings and are cute and fun to boot. Some are actually just normal varieties picked while still small and immature (such as baby zucchini) or are the secondary fruit that ripens after the first and larger crops (tomatoes get progressively smaller as the season progresses). Other mini varieties of fruits and vegetables are actually selectively hybridized for small-sized produce. These miniatures are every bit as nutritious and tasty as their full-sized counterparts, just in smaller packages; in fact, sometimes the flavors can be more intense and the textures more defined. You might just want to grow these for your own family, friends, or neighbors because they're fun and tasty, but if you're contemplating launching a community-supported agriculture (CSA) operation or another way to bring in money from your efforts, baby fruits and veggies could be worth looking into.

They first began showing up in the public eye as garnishes and in side dishes and salads at fancier restaurants, and they've caught on with the general public. Smaller vegetables and fruits—for example, watermelons—have a practical appeal as well. They take up less room in the refrigerator

Baby corn is a popular ingredient in some Asian dishes.

and are more appropriately sized for small families or singles. In addition, individuals who like to entertain find these baby veggies and fruits—well, entertaining! Baby greens are increasingly sought as a convenient ready-to-eat product, and the University of Kentucky Cooperative Extension Service states that market options for baby vegetable producers include sales to eating establishments, particularly high-end restaurants. Gourmet and health-food markets, as well as local specialty food stores, offer potential sales, as do some caterers and food-service establishments.

Authentic baby carrots look like carrots in miniature, not simply cut and shaped like small carrots.

If not from a midget varietal, baby veggies are picked in an immature state and are more perishable than when harvested at maturity. Therefore, they don't ship well and have a shorter shelf life, a fact that some hobby farmers and CSA operators use to their advantage—they can provide fresh baby vegetables to local markets at competitive prices, while out-of-state producers must pack and ship, a costly process that they pass on to their customers.

Other than spacing and harvesting (miniature varieties take up less garden space), most miniature vegetables/fruits have the same soil, water, nutrient, and light requirements as their full-sized counterparts. Be aware that, as cute as they are, it takes a whole lot of midget veggies to provide the equivalent of a full-sized vegetable of the same variety. If you're planning on freezing or canning produce, you'll most likely want to plant normal-sized varietals instead of or in addition to miniatures. That way, for most plants, you can harvest both midget and mature produce according to your needs and desires. Tiny green beans are oh-so-tender, but if you let them grow, they'll fill your canning jars with full-sized beans. Leaf crops such as lettuce, spinach, and kale can be sown more densely if you plan on cutting them in the immature stage before thinning becomes a necessity. Miniature, midget, and dwarf vegetables also grow well in containers because they need little above-soil growing space, and normal growing practices apply. Early harvest may make pest control more challenging because you don't want to apply deterrents, even organic ones, too close to harvest time, but good cultural and pest-management practices will lend success to your venture.

Baby vegetables, like all babies, grow up almost overnight, so you'll need to keep an eye on them to harvest at the desired stage. Handle them carefully, especially those miniatures that are immature stages of a larger variety, because their skin and leaves are more tender and more easily bruised than when full grown. Certain crops, such as baby sweet corn, can be more labor-intensive to harvest. Because of that, you can ask a higher price if you raise them to sell. The University of Kentucky Extension mentions that microgreens can sell for close to five times as much per pound as normal-sized greens. According to Texas A & M's Agrilife Extension, there are about 45–50 types of baby/midget cultivars currently marketed in the United States.

Miniature Vegetables

Baby artichokes:	These have no "choke" (center inedible core). Once the outside leaves are peeled, the whole vegetable is edible.
Baby beets:	Ranging in size from a golf ball down to a quarter, gold varieties have a milder, sweeter flavor than reds.
Baby carrots:	Not to be confused with "manufactured" baby carrots, which are really pieces of regular-sized carrots milled to baby carrot size, true midget varieties will look exactly like full-sized carrots, but small. They are very sweet, and most varieties can be eaten greens and all. They are available in various colors, including white, orange, and red, and range from approximately 1½–5 inches long.
Baby cauliflower:	Baby snowball cauliflower is 2 inches in diameter and is similar in taste to the mature variety.
Baby celery:	Harvested when about 7 inches long, it tastes stronger than the mature version.
Baby corn:	Available in white or yellow varieties, baby corn is harvested just after the silks emerge.
Baby eggplant:	Some varieties can be bitter and contain many seeds.
Baby French beans:	Commonly called *haricots verts*, this is a small, flavorful strain of green bean that was developed in France.
Baby green onion:	Its mild taste is similar to that of chives.
Baby lettuce:	Varieties such as royal oak leaf, romaine, green leaf, and iceberg are cut at the optimal size of 4–6 inches.
Baby melons:	Watermelons, cantaloupe, and honeydew are true miniatures; full-sized varieties picked small would not be ripe. Miniatures range from 3–13 pounds in weight and are sweet, tasty, and easy to fit in your refrigerator.
Baby potatoes:	There are miniature cultivars that produce potatoes 4–5 inches long at maturity.
Baby pumpkins:	They are very cute and can range from 2–4 inches in diameter.
Baby scallopini squash:	This summer squash comes in green and yellow and tastes reminiscent of larger relatives.
Baby soft-shelled squash:	Immature squash of regular-sized varieties, including zucchini, pattypan, and crookneck, these are often sold and eaten with the flowers still intact.
Baby tomatoes:	Cherry tomatoes are a popular baby variety, and baby tomatoes are available in a multitude of shapes and colors: the red and yellow varieties are popular, and the latter tend to be quite sweet.
Baby winter squash:	This is a miniature acorn-squash variety that can be picked when the size of a golf ball.

Bright yellow lemon cucumbers and pattypan squash are interesting varieties on more common favorites.

Be Different

Different is in. That's the bottom line. Why go with what we already know, right? This adage is especially true in the realm of foods as consumers look for something new to add pizzazz to their plates. So color is definitely in. Purple, cheddar-orange, and yellow cauliflower. Carrots ranging from creamy white to yellow, orange, and deep red. Red sweet corn. Swiss chard with its dark green leaves and bright red veins. Kale in its various sizes and colors from bronze to purple. Why grow "same 'ol same 'ol" when flowering cabbage comes in white and fuchsia, and potatoes can be deep purplish-blue or as golden as butter? You can even find "pink" pumpkins (their flesh is bright orange inside). Whether it's just for your own table or for your customers at the farmers' market, colored veggies are in demand.

Different shapes and flavors are catching the public's eye as well. Right at the top of the list is Witch Finger grape. A purple-red cultivar made by crossing certain varieties of wild grapes with a sweeter domestic strain, the fruit is narrow and elongated. Another grape that's in demand is Cotton Candy—a green seedless variety whose taste and smell give rise to its name. And then there's the cross between Russian red kale and conventional Brussels sprouts marketed under the brand names Lollipops and Kalettes. Eaten raw or cooked, the leaves are like miniature kale leaves but more tender and lacking the thick stem in the middle. Firmer centers have more of the nutty, sweet Brussels sprouts flavor. At this point, this hybrid is hard to come by and not very easy to grow, but those who have tasted it report that the mix of soft and hard textures makes the colorful vegetable ideal both raw and cooked in all sorts of ways.

The apricot, an old favorite, is experiencing a resurgence of popularity, and several new varieties, including a red apricot, have fueled this comeback. Meanwhile, the ancient and lowly leek, a favorite of Aristotle and the emperor Nero, is being touted by a whole new generation of cooks and foodies. With its subtle onion-like flavor, this member of the allium family is not difficult to grow and lends itself to a multitude of culinary dishes. And then there are cucumbers. These perennial favorites are receiving new interest in such varieties as mini-cucumbers, which are the size of a penny; and lemon cucumbers, so-called because they have the size, color, and rounded shape of lemons. The latter, an heirloom variety, came on the scene in 1894, but it tastes and feels like a modern discovery: sweet, crunchy, and easy to grow.

Branching Out

Vegetables, fruits, and berries are showing up in the wonderful world of landscaping, too. Often called "edible landscaping," integrating food plants into decorative settings in place of straight ornamentals is growing in popularity. It not only enhances gardens with aesthetically appealing components, it just makes sense. Why grow an ornamental shrub when you can plant a blueberry bush and have visual interest, color changes, and edible fruit? Edible landscaping mixes beauty, utility, and sustainability. Purple beans are finding backyard fences a great place to trellis and show off their stuff, and it's amazing how showy certain varieties of cabbage and kale can be when placed right smack in the middle of a flower bed. A veggie tucked in next to a flower, a goji bush splashing its bright red fruit against a weathered wooden fence—both aesthetically and gastronomically, the entrance of edibles into what traditionally has been solely decorative landscape is a great way to introduce the sustainable victory garden concept without losing the beauty and color we strive for in our outdoor spaces. Plan carefully and enjoy the serendipity of finding food in unusual places.

A coral-bark Japanese maple against a backdrop of snow.

Speaking of beauty, color, and departing from the norm, foliage is on the rise as well. People still love flowers, but foliage plants are gaining admirers of their own. The colors range from frosty gray-green to red-tipped and rosy. In addition to their lovely hues, gardeners are discovering that foliage plants—bushes, trees, grasses—offer wildly divergent shapes and textures, lending new life and interest to yard and landscape. An added advantage, especially in regions where flowers don't bloom in the winter, is that you can have color and form in your backyard year-round. Bushes, grasses, shrubs, and trees may go through seasonal changes, but often those changes introduce new color or texture to your scene at times when flowers are nonexistent.

One example is the coral-bark variety of Japanese maple, *Sango kaku*. New growth for this small deciduous tree splashes the spring garden scene with glowing buttery yellow, which transitions in summer to a bright and grassy green. Come fall, those leaves begin to return to their golden beginnings, sometimes taking on a luminescent hint of orange as autumn progresses. When the leaves let go of their branches, we find that they've been hiding bark that has turned from the nondescript gray-brown of summer to a brilliant coral red. Four seasons, four lovely experiences, one tree. Ornamental grasses put on a real show as well, adding color, texture, and movement to a static corner or hiding a utilitarian feature such as a water or gas meter.

Interest in dwarf trees, weeping trees, bushes, and trees with unusual bark (such as river birches) or branches (think corkscrew tree) are on the rise as well. A good example is the manzanita, a native shrub ubiquitous to Northern California that adds color and form, as well as tenacity, to dryer climates. Much like the kaleidoscope eucalyptus of the Philippines, the manzanita's new growth borders on chartreuse. Parts of the branches retain their rich burgundy color from the previous year, often with a twist of silver-gray spiraling up, showing where the branch is no longer growing and contributing to the twisted formations. As green fades to burgundy, little bits of "bark" fleck the surface, but if you run your hand down the branch or trunk, those little flecks fall off, revealing the

A stone path lends an elegant touch.

burgundy satin-smoothness of the new bark. Rounded, silver-green leaves a bit reminiscent of eucalyptus (although without the characteristic smell) remain year-round.

The echoing of color or texture in miniature or with slight variations ties the landscape together without making it boringly repetitive. Perhaps this "year-round-ness" and symbiosis of foliage coupled with an increased focus on sustainability has given rise to yet another trend: perennials. This does, of course lend a certain challenge to achieving a constant play of color throughout the growing season, but flower gardeners are finding new satisfaction in old favorites, learning which perennials bloom when and getting in touch with the old and charming practice of *friendship gardening*—the gifting and receiving of plants (usually perennials) from relatives, friends, or neighbors. "This is what I call Norma's columbine, and these are Granny's peach-colored iris. Here in this corner is my great-grandmother's rose—from the bush she always kept blooming by her porch swing in Kentucky. I'm starting a bush for each of my daughters and my sisters."

Dressing Up Your Yard

Gardeners rural and urban are returning to the land in multiple ways, including by decorating their gardens and landscapes as living spaces. This goes beyond patios. We're talking about such things as creating little (or big) areas in which to relax, entertain, cook, eat, and even sleep. Picture a hammock strung under the birch trees with a small table on which to set your iced tea (or homegrown green juice) while a fountain trickles just close enough to waft you away to another time and place—creative outdoor spaces can be magical, and people are turning their imaginations loose in their yards. If you want to get some inspiration from other people's ideas, visit Pinterest (www.pinterest.com) and Houzz (www.houzz.com). Here is a sampling of some of the more popular elements with which gardeners and nature dwellers are decorating their outdoor spaces.

- An old porch swing with comfy, colorful cushions hung under a big old tree and, above it, a shabby-chic vintage chandelier. It makes a statement with no wiring, but with a little extra time, you could make it functional.
- A cozy dining spot under an arbor, with LED lights woven through the grapevines or honeysuckle, rustic-wood or wrought-iron French terrasse chairs, and a canopy of leaves overhead.
- A curving path of bark, river gravel, slate, or other material meandering through the landscape to a reading bench nestled among pungent herbs.
- Accent cushions to visually pull the seating area into the garden and the garden into the seating area.

- An outdoor cooking area that will have you wondering whether you're outside or in until you look up and see the sky.
- A vintage window-turned-shelf hung on the fence with a vase in which you can place the fresh flower of the day.
- A rustic mirror reflecting a show-stopping plant.
- A wandering stream complete with waterfalls and a pool, with a redwood bench or cozy chair nearby.

Containers are finding popularity among vegetable gardeners.

The rise of outdoor decorating, lawn games, and garden parties is spiking the demand for garden furniture, fountains, outdoor chandeliers, decorative planters, bird feeders, candleholders, and more. Along with outdoor decorating, geometric themes (circles, triangles, squares) are popular as gardeners are straying away from more formal straight lines.

Color is integral to backyards, flower beds, and landscaping. Consider a monochromatic color scheme; if you have a favorite hue, now's the time to showcase it, letting foliage and flowers echo each other in every corner.

Container Gardening

Container gardening is trending, even among people who have plenty of space in which to garden. Combining flowers, herbs, and vegetables in the same containers is also gaining in popularity. For one thing, the beauty and variety of pots and containers is mind-boggling in color, texture, shape, and size. Used as accents, in groupings, or in a garden in their own right, containers are winning fans in cities and on the farm for their ease and instant beauty.

Container planting basics include:

- Choose the right container for the job—larger pots give roots room to grow, resulting in lusher top growth. Interestingly, larger pots also require less water, and your plants will be at less risk of a sudden drought if you forget to water. Also, plants can stay potted longer in larger pots, saving you the labor of having to relocate them as they grow.
- Be sure that your containers have drainage holes in the bottom. If not, plants will drown. The exception is pots designed to be self-watering.
- Use a good potting mix to help plants grow larger and live longer. Do not use topsoil or garden soil; even soil sold for containers tends to be much too heavy, retaining too much water and leading to root rot and untimely plant death.
- Space plants closer together in containers than if you were planting in the garden. Wet the root balls and snuggle them up right next to each other to get the most attractive results right away.
- The right fertilizer is important, and, unfortunately, it's easy to overdo or underdo. One fertilizer that many container plants do well with is Dynamite slow-release fertilizer, which is recommended by container planting expert Pamela Crawford.

Herbs alone and with flowers make great container plants, as do vegetables, especially the miniature varieties mentioned earlier. You'll have the added bonus of harvesting when you want extra-fresh ingredients.

Farm Animals

A farm hardly seems like a farm without some animals. For many people—including some hobby farmers—animals are mostly companions, and this is enough in itself. There is no rule that says that you can't raise or own an animal merely to give you pleasure. For example, although farmers sometimes use horses for work or raise them for profit, a horse's predominant role on the farm tends to be as a recreational pastime and/or a majestic lawn/field ornament. Something about gazing into the dark brown depths of a horse's eyes or watching two or three charge across an open field draws time to a halt. Even chickens, useful as they are, can double as farmyard therapists. Leaning against a tree, listening to the contented clucking of hens as they scratch for whatever they may find, is rarefied treatment indeed.

Most farmers understand that while animals are comforting, companionable, and enjoyable, they are also a means of food and products: in exchange for a bit of food and shelter, a few chickens can contribute eggs for the family. A cow or a goat can provide milk and meat. Sheep and goats can help control weeds and supplement the yearly meat supply, and their wool/hair can be used in handcrafts.

If you didn't grow up on a farm, it can be hard to understand how this year's batch of darling little chickies can be dispatched for next winter's chicken soup, or how ol' Rosie's rollicking steer calf could be slated for the freezer next year. But this is the reality that farmers hold in balance, and those who raise animals for commercial purposes understand this as well.

A young Holstein calf.

If you have small children and plan on "multipurpose" livestock, *Charlotte's Web,* E. B. White's beloved childhood book, is not going to be the best reading material to help them deal with this aspect of farm life! It's best to talk about this as a matter of course, so that they grow to expect this year's pigs, calves, chickens, and the like to give place to next year's crop. Many families have helped lessen their young ones' sadness by making sure that certain long-term animals—cats, dogs, horses, milk cow or goat—are named and celebrated, but they tend to let the piglets, chickens, turkeys, and calves go unnamed and unsocialized, merely fed and cared for, but not bonded with. In the case of turkeys, chickens, and pigs, the animals themselves help in this regard. Wilbur in *Charlotte's Web* may have been one impressive porker, but, in reality, as a woman who grew up on a small Montana farm said, "I'm sorry to break the ideal picture, but pigs just plain stink. When you want to move them anywhere, they squeal. The ones I fed growing up were forever breaking out of their pens and we had to chase them for ages to get them back in. And they'll run right over you if you're in the way."

One hobby-farm kid in California had this to say about chickens: "They're dumb. They smell, and I'll never, *never* make my kids take care of them." But that same youngster, who helped in the yearly chicken harvest, would sit for hours on the back of her pony, reading books and just enjoying life.

To do justice to the subject of farm animals and animal husbandry would take at least one entire book, if not several, but perhaps this chapter will serve as an overview and can steer you toward delving deeper into whichever areas you desire to explore further.

The Menagerie

The farmyard is a cacophonous, wild, and joyous place. You may be greeted by the neigh of a horse or the bray of a donkey, the gentle lowing of a mother cow talking to her calf, or the hissing of a gaggle of geese warning a visitor. A llama may guard the flock of sheep that are playing king-of-the-hill on a pile of manure. Life is abundant and entertaining. What kind of animals do you want? The possibilities are wide open.

Horses, Chickens...Llamas?

Horses are often the first animals that new farmers get, as a result of either their children's pleading or their own dreams. Horses have worked for people for thousands of years—providing power and yielding progress. Horses are expensive pets, but they can earn their keep by providing recreation or working in the field and forest.

Chickens are almost a must-have farm animal. They are economical, supplying meat and eggs that taste so much better than grocery-store fare, and they require little effort. Chicken meat and eggs from small flocks are relatively easy to market. Other members of the bird clan include turkeys (which are smarter than most people give them credit for), guinea fowl, pigeons, peacocks, ducks, and geese. You can often purchase baby birds at feed stores in the spring, or you can order them from hatcheries, which ship them to you as day-old chicks (or poults or ducklings). It is always a great day when your baby birds arrive. You can also purchase your own incubator and eggs to incubate, or, from time to time, you may find someone willing to sell a few adults from his or her flock.

Need it be said that cows are wonderful animals? Beef cattle require no housing unless calves will be born in the winter, and dairy animals do fine with minimal housing. For a homestead approach, one dual-purpose animal, such as a Jersey cow, can raise a calf and supply milk for the family.

With machinery available to do much of a horse's work, many hobby farmers enjoy their horses simply for recreation.

Children are often fascinated by hatching and raising chicks.

Pigs have personality plus. They are very intelligent and entertaining, but they can also be challenging. They are very curious, so they often get into things that you don't necessarily want them into (the garden, for example). And, since they are omnivores, they may eat things you don't want them to consume (such as one of your chickens).

Goats are the magicians of the farmstead—able to escape from enclosures like the legendary Houdini. Contrary to popular myth, they don't eat tin cans, but they can wreak havoc on things. For example, given the opportunity, they will climb all over your vehicles, leaving dainty little hoof dents on the roof, the hood, or the trunk. However, goat milk is very healthy, and goat meat has good market potential, especially in certain ethnic communities.

Sheep are sweet-tempered and vulnerable but can also high-strung, making them somewhat difficult to handle. They tend to move like a school of fish, in a tight bunch, so working with them requires patience. Yet it is inexpensive to start sheep; you can raise them in a small area and build an impressive-sized flock pretty quickly from a few animals. Research genetics and talk to sheep farmers in your area who have already ascertained which breeds do best in your part of the country. It matters—sheep that are of more tropical descent will not thrive in the colder regions.

Llamas and alpacas are members of the camel family. Farmers raise them primarily for their fleece, which hand-spinners and weavers crave, although you can also train them as pack animals. Llamas also make good guardian animals for sheep and great conversation-starters.

Shopping for Animals

Buying your first animals is an adventure. Unless animal husbandry and farm life is in your background, you'll need to learn about markets, conformation, disposition, and the like. Know what you want to buy and how important it is to you that the animal you're going to see matches that ideal,

Pigs can be a lot of fun but can be challenging to care for.

because it's difficult to look into an animal's eyes and say no. Another caution: Unless you've had experience or you have a neighboring experienced farmer who can accompany you, you might want to avoid sale barns. Although many animals that go through the barns are healthy, some are being sold because they have failed to breed, because they are older, or because of some other aspect that may make them a less-than-optimal choice

Goats are popular and personable members of the hobby-farm menagerie.

for your farm. In most cases, you will probably have the best luck buying through a "private treaty" deal, which means directly from the animal's current owner instead of at an auction or through a broker. Owners selling directly are more likely to be honest about the animal's disposition and health, but remember that when buying any animal from anyone, there is a certain amount of "buyer beware"—once in a while, you may find an owner who isn't totally above board about an animal's qualities.

Some of the best advice I've gotten about going to view animals is to arrive anywhere from five minutes to an hour earlier (apologizing profusely, of course!) than the seller is expecting you. Mark Rashid, the author of three of my favorite horse-related books, including *A Good Horse is Never a Bad Color* (Johnson Printing, 1996), makes the following observation: "By 'accidentally' arriving early when looking to buy a horse, I've seen supposedly quiet and gentle horses that kicked when you walked behind them, bit or bolted when you tried to cinch them, pulled back when you tried to

tie them, and threw their heads when you tried to bridle them. In one case, I even walked in on a fellow as he was injecting the horse he wanted to sell me with a sedative."

If you have a friend or acquaintance familiar with the type of animals you are looking at, ask him or her to go with you. That person will be able to help you evaluate the animals' health, condition, and suitability for your needs. Also, if you are buying expensive animals (breeding stock, show animals), I strongly

Alpacas, close relatives of llamas, are prized for their beautiful fur and gentle personalities.

Horses are herbivores that obtain a portion of their feed from grazing.

recommend having a veterinarian check out the animals before you close the deal. The money you spend on a prepurchase exam could save you not only dollars but also heartache down the line.

Feeding Animals

One of the most important roles you will play in the lives of your animals is meeting their nutritional needs. The majority of farmyard animals are herbivores, so plants should be the exclusive source of their nutrients. Pigs and poultry are omnivores, so at least a portion of their natural diet is composed of nonvegetable protein.

Even within species, food needs vary with the animals' ages, the amount of work or production they perform, and the seasons. As anyone who has owned animals great or small knows, you must develop an eye for their condition.

The good news is that feeding your animals properly does not take a degree in veterinary science. It's a skill acquired naturally as, day after day, you check on your animals. Seeing them whole and healthy is the best way to educate yourself, so that if one day you see something out of the ordinary, you'll be able to spot it immediately. And another thing: Although you'll need to adjust feeding regimens according to the ever-changing needs of your stock, there's no need to discover through trial and error what best keeps your animals in healthy condition. Expert advice awaits through a myriad of avenues: extension offices, veterinarians, farming neighbors, breed specialists, dedicated websites, and more. Nonetheless, here's a primer.

Maintenance rations refer to the amount of food required to support the animal when it is doing no work and producing no product (milk, meat, eggs, fiber). A maintenance ration basically supports all minimum body functions, such as respiration and cardiovascular function, holding the animal's weight steady. For most farm animals, up to half of their food ration represents their maintenance requirement.

Providing adequate feed for growth, reproduction, and/or work is really the overarching goal of every livestock owner. And *adequate* is a key concept in my mind— you want to feed enough to keep your animals healthy and strong, but you don't want to overfeed them. Overfeeding not only costs more than is necessary, it isn't good for the animal. An overly fat breeding animal has trouble breeding. If you feed young animals too much too quickly, they can develop joint and bone problems from growing too fast. This is especially true for young horses and puppies of the larger dog breeds.

Pigs are fast-growing animals but should not be overfed.

Feed Components

Whether the source of feed is plant or animal, there are certain basic properties that all feedstuffs share. Three major components make up feed: water, organic matter, and mineral matter (ash). Carbon, hydrogen, and oxygen make up the organic matter, although other elements may be present. You can further divide organic matter into four categories: carbohydrates, fats and fatty substances, nitrogenous compounds (proteins), and vitamins.

Carbohydrates, made up of carbon atoms attached to water molecules, can be further broken down into sugars, starches, and fiber. The proportion of these components varies according to the type of plant, its age, and environmental factors, such as drought. The sugars and starches are easy to digest, so they provide a relatively high feed value. Lignins primarily make up the fiber component and are completely indigestible. Cellulose, which accounts for about 50 percent of the organic carbon on earth, is also present in fiber. Cellulose requires bacterial fermentation to break it down into usable sugars and starches. Although many animals are able to ferment small amounts of cellulose in their intestines, only ruminants such as cattle and sheep can convert the bulk of the cellulose in their diet into usable sugars and starches.

As with carbohydrates, carbon, hydrogen, and oxygen make up fats and fatty substances, although in different proportions than in sugars and starches. In spite of our current preoccupation with nonfat products, fat is an essential nutrient, especially for young animals, providing more than twice the energy that a carbohydrate does; it also helps animals maintain body condition and temperature. In addition, some vitamins are fat soluble, and without a certain amount of fat to carry them to the cells, the body cannot absorb them, which leads to vitamin deficiency and health issues.

Just as humans do, animals need protein for the development of all cell walls and for forming muscles, internal organs, blood cells, hair, horns, and bones. In most cases, protein accounts for 15–20 percent of the animal's weight. Unlike sugars, which may contain as few as twenty atoms, thousands of atoms make

Feeding requirements differ for free-range chickens and those kept in confined spaces.

up each molecule of protein. In nature, as in construction, it's often easier to build a complex structure by using substructures, like building blocks. In the case of proteins, amino acids are the building blocks that simplify construction, and there are about twenty critical amino acids.

Vitamins are also organic in nature and required in very small quantities. Just like humans, animals can experience a wide range of diseases, such as rickets and anemia, resulting from vitamin deficiencies. At the same time, some vitamins that accumulate rather than filter out of the body when in excess (vitamin A is one of these) can be toxic if given in too high a quantity.

And then there are minerals—elements such as sodium, calcium, phosphorous, and selenium, to name a few. As with vitamins, most minerals aren't required in large quantities; toxicity can occur from excess minerals, while deficiencies can cause a wide range of health problems. Mineral deficiencies (and excesses) usually occur where soil mineral imbalances exist. Plants grown in soil that is too low or too high in any given mineral will carry that profile, which, in turn, will be reflected in your animal's tissue. In addition to testing your soil (see Chapter 5), ask your extension agents, a reputable feed dealer, or your veterinarian to provide you with information on soil mineralization in your area and recommend the possible mineral supplements needed for your situation.

Feeding Babies

It's no surprise that the best feed for baby animals comes right from their mothers (unless you're talking poultry, of course!). At times, though, farmers need to step in and hand-feed little ones, either because the mother is ill or dead or, as is often the case with milk cows, because the bulk of the milk goes for human consumption. Sometimes, a mother won't accept a calf or lamb; other times, the baby is too sick or weak to suckle the mother.

In commercial dairy production, almost all calves (or kids) are bottle-fed. Bottle-fed or not, it is crucial that all newborns receive *colostrum* within the first twenty-four hours of life. Colostrum is the first milk that a mammal secretes after it gives birth, and it jump starts the baby's immune system. The University of Minnesota Extension states that only during the first twenty-four hours can newborns absorb whole antibodies through the walls of the small intestine, and these antibodies stay in their systems for weeks. Morbidity (illness) and mortality rates for calves who do not receive colostrum during this crucial window are more than double that of newborn calves who received colostrum.

Milk replacer is available for baby animals who must be bottle-fed.

If, for some reason, the mother's colostrum isn't available (as with an orphaned animal), there are commercial colostrum products that you can purchase from your vet. If you have planned ahead, you can use colostrum saved (and frozen) from other animals in your herd. It is best if colostrum comes from the same species, but, in a real pinch, you can use cow's colostrum for other species. Most dairy farmers keep frozen colostrum, which they will usually share in an emergency.

When bottle-feeding, the rule of thumb is to provide 10 percent of the baby animal's body weight per day in whole milk, preferably from their own species. However, goat's milk or cow's milk will work for most babies. If milk is unavailable, you can always use commercial milk replacer. Manufacturers make replacer for most classes of livestock (and even for dogs and cats). Calf milk replacer is readily available from feed stores, but you may need to special-order replacers for other species. Look for unmedicated milk replacer. As Dr. C. E. Spaulding says in *A Veterinary Guide for Animal Owners*, "There is not enough antibiotic in a pound of feed [medicated milk replacers] to prevent scours or other diseases, and the antibiotic fed daily can damage enough of the natural and necessary bacteria in the gut to cause scours." In addition, look for milk replacer that lists milk as the first ingredient on the ingredient list; cheap brands often include no milk at all.

Furthermore, don't feed babies more than they are supposed to have just because they act as if they're still starving after a feeding. If possible, feed them smaller, more frequent milk rations throughout the day rather than two big feedings. A schedule of frequent feedings mimics what the baby animals would do if they were able to access their mothers throughout the day. Bottles, nipples, and other feeding accessories are readily available at local feed or farm supply stores and, of course, through online retailers.

If a herbivorous baby isn't "housed" on pasture, offer a small pile of hay for him to nibble at right away. However, realize that it takes a while for a newborn's gut to build the healthy bacteria

A ten-day-old goat drinks milk from a bottle.

that enables him to extract nutrients from graze. Foals (baby horses) have very little microbial activity in their guts before three months of age, and ruminants can take up to six months to develop their rumens. Meanwhile, they need milk to grow. When baby herbivores are out in the pasture alongside their mothers, they mimic the adults' grazing. As they introduce small bits of forage to their diets, their digestive systems develop to maximum efficiency, reaching the point at which they no longer need milk.

If you're raising a ruminant baby on a bottle, it can be helpful to supplement milk with hay-based pellets that contribute to the microbial growth in the rumen. As previously mentioned, calf starter is available at most feed stores, but many commercial products may contain antibiotics. While this sounds like a great idea on the surface, in reality, antibiotics can inhibit the growth of the good digestive bacteria while actually not delivering sufficient protection from disease and infection.

You may have to teach a young animal to eat supplements. After feeding him his milk ration, and while he's still smacking his lips together, place a small handful of pellets, oats, or another feed of choice in his mouth. He may dribble out more than he swallows in the first day or two, but give him free access to his ration, and he'll soon figure it out.

Digestion

We classify animals according to the configuration of their digestive systems.

- **Monogastric:** Possessing a single stomach (i.e., pigs, poultry, and humans)
- **Postgastric fermenter:** Possessing a single stomach but a well-developed cecum, a fermentation chamber located between the small and large intestines (horses and rabbits fall into this category)
- **Pregastric fermenter or ruminant:** Possessing a four-compartment stomach (not four stomachs, as is commonly thought), including the rumen, or first compartment, which acts as a large fermentation chamber. Cattle, sheep, and goats are common farm ruminants. Pseudoruminants, such as llamas and alpacas, are a smaller class of pregastric fermenters. The pseudoruminants have three stomach compartments instead of four, but their digestive process is very similar to that of true ruminants.

Animals with monogastric digestive systems have some disadvantages when it comes to digesting food. Because they can ferment only a very small amount of the fiber in their diets, they can synthesize only a few of the many amino acids that their bodies require from vegetative matter. This means that they get almost no feed value out of hay or straw; however, young vegetative green grass can provide them a small amount of feed value. Monogastric animals (e.g., pigs, cats, dogs) thus need a wide variety of protein sources in their diets to meet their amino-acid needs. Although they can't use grass to supply amino acids, monogastric critters have teeth and stomach enzymes that allow them to break down animal or insect-based protein sources.

The postgastric fermenter's cecum performs a similar function to a cow's or sheep's rumen. Although all mammals have this organ, it is only well developed and highly effective in postgastric fermenters, such as horses and rabbits. The cecum, located between the stomach and the small intestine, has one opening. Food begins enzymatic digestion in the stomach and then passes into the cecum to undergo fermentation via bacteria and microbes, which further break down the fiber and extract the nutrients before the food empties into the large and small intestines. Here, the food undergoes continued fermentation while water is largely reabsorbed into the system and waste products are filtered in.

Some postgastric animals—horses are an example—do not do as well on two big meals a day with lots of empty stomach time in between. Because they are grazing animals, their stomachs produce powerful acids 24/7, and an empty stomach can lead to ulcers. According to the Ontario Ministry of Agriculture, Food, and Rural Affairs, it takes an average of 1 ½ hours for a horse's stomach to empty, so keeping grass or hay/forage available gives horses the fiber they need to absorb the acid and keep their guts healthy. To avoid overfeeding, some horse owners use a mesh hay bag to limit consumption—it takes a horse a lot longer to pull the strands of hay through the mesh, keeping it from overeating while at the same time helping to keep food in its digestive tract.

When it comes to fiber digestion, ruminants are the clear winners. The University of Wisconsin Cooperative Extension states that the rumen contains billions of bacteria and that 1 milliliter of rumen fluid contains 10–50 billion microbes and more than 1 million protozoa, which attach to the

Sheep belong to the ruminant category.

Ruminants graze to form "cuds," which they chew thoroughly to break down and synthesize nutrients.

feed particles to digest the feed. The rumen, where fermentation begins, is the first compartment of the stomach. Ruminants graze, cropping forage and chewing only briefly to moisten and form mouthful-sized wads called *cuds*.

When the rumen is full of these cuds—a fully grown cow can hold more than 35–40 gallons!—or the cow has munched through its morning hay supply, it will regurgitate each cud (which has been soaking in all of those digestive bacteria and microbes) and chew each one more fully. Ruminants can spend up to six hours per day eating and up to eight hours per day "chewing cud," or ruminating. Cud-chewing serves two purposes: it provides some extra mechanical breakdown of food, and it adds lots of saliva to the rumen, further aiding in the digestive process.

After swallowing, the chewed cud passes easily to the second stomach compartment, the *reticulum*. This chamber performs two functions: first, it traps large feed particles—whole grains and the occasional rock or other foreign object—so that they do not proceed to the later digestive processes without being broken down sufficiently; second, it collects material for further rumination (process of breaking down foodstuffs into usable particles). The third compartment, the *omasum*, serves as a filter, squeezing water from the feed so that the majority of it stays in the rumen rather than passing on into the digestive tract. After the omasum is the *abomasum*—*ab* meaning "from, off, or away from." The abomasum performs the same basic role as a monogastric's stomach: adding buffers, acids, and enzymes before passing the contents on to the small intestine and the rest of the digestive process. Thus, ruminants are able to fully break down the fiber they eat and synthesize all necessary amino acids from whatever nitrogenous compounds are present in the vegetation.

A word about feeding grain: Grain that bypasses the rumen generally remains intact, which is why you'll see grain kernels in the manure if you're feeding whole grains to your ruminant livestock.

Change in Diet

An important objective of any livestock feeding program, ruminant or otherwise, is maintaining an environment that is good for the digestive system's microorganisms so that they can do their jobs well. One component of this is to make any dietary changes slowly—say, over the course of two weeks—so that the digestive flora can adjust to the change.

Those whole kernels not only cost you money, but they don't benefit your animals because the nutrients are not extracted from them if they pass directly into the reticulum. Try feeding ground, rolled, or cracked grain, and only feed whole grains with hay or other light fibrous feeds so that the rumen captures the whole grains and begins to digest them before they move on to the reticulum.

That said, a ruminant's system is not designed to process grains. Concentrated

A native grass to Iowa, switchgrass offers abundant pasture in the summer, when other grasses are less productive.

feeds acidifies their systems and can lead to health issues for the animals and for the consumers of animal products—us. Thus, meat from grass or forage-fed animals is healthier for consumers than their grain-fed counterparts, and forage-based ruminants are much healthier overall than grain-fed animals. For a country raised on grain-fed meats, the taste of grass-fed meats may take a bit of getting used to. Also, grain/corn-finished meats tend to be more tender and more evenly marbled than forage-only meats, but they are also higher in saturated fats and lower in omega-3s.

One side effect of fermentation in a ruminant is that it produces a significant amount of methane gas. In fact, a mature cow can produce up to twenty cubic feet of gas per hour, which is enough gas to fill a balloon the size of a large chest freezer. The cow passes this gas through regular belching (eructation).

Quality Pastures

Livestock are capable of turning grasses into a high-quality protein for human consumption, making pasture a critical component of livestock operations. If they have access to pasture or hay, most animals (pre- and postgastric fermenters) do fine with no supplemental concentrate. When hay or pasture is inferior, your animals may need a concentrated supplement. Dairy animals and those working hard (pulling a plow or being ridden for hours each day) may also require supplemental concentrates. In addition, pregnant, lactating, and breeding animals may need supplementation if the pasture or hay isn't at its highest quality.

As farmers continue to move toward sustainable agriculture and animal husbandry as well as healthy food production, one aspect that is receiving attention is *grazing genetics*, especially in milking breeds and increasingly in sheep. Good grazing genetics produce animals that thrive on a forage-only diet without the supplement of concentrated feeds. While grazing genetics are most commonly discussed regarding cattle, Ulf Kintzel of White Clover Sheep Farm (www.whitecloversheepfarm.com) in the Finger Lakes region of New York raises White Dorper breeding stock and freezer lambs entirely on grass, and he is part of a growing constituency. So if grass-fed, sustainable livestock is a goal for

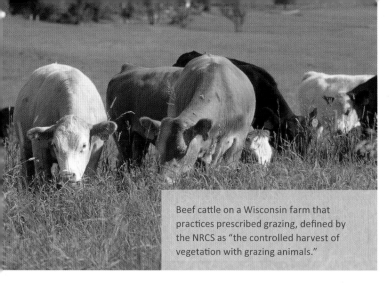

Beef cattle on a Wisconsin farm that practices prescribed grazing, defined by the NRCS as "the controlled harvest of vegetation with grazing animals."

you, be aware that some breeds are not readily able to maintain good condition on pasture only, and, as with so many other aspects of farming, do your research before purchasing animals so that you start with the best potential for success.

Depending on your part of the country, you might notice that where there is livestock, there is often a lot of bare dirt or weeds in pastures. But the good news is that with a technique known as *managed grazing*, you can take excellent care of both your animals and your land. With just a little extra money and time spent initially in developing a subdivided pasture, you will reap benefits year-round, including:

- Good grass cover, which is aesthetically pleasing and increases your property value
- Reduced feed bills and reduced weed pressure
- Reduced animal health problems, such as those related to parasites and dust
- Reduced wind and water erosion and nutrient pollution (the Environmental Protection Agency [EPA] recognizes the technique of subdividing your pastures as environmentally beneficial)

In a healthy pasture, there's a complex and diverse group of animals and plants that interact with each other. One goal of managed grazing is to foster this healthy complexity. The variety of plants, animals, insects, and microscopic organisms that inhabit a healthy pasture differ regionally. But, wherever you live, if the pasture is healthy, many creatures call it home.

Ideally, a pasture should contain about 60 percent grass and 40 percent legumes. If yours doesn't have enough legumes, talk to your local county extension agent to find out what types grow well in your area. Introduce legumes simply by spreading seed in early spring on top of the ground, either with a "whirlybird"-type seeder or by throwing handfuls out as you walk around. If you have an ATV, you can buy a seeding attachment that makes sowing large areas less daunting.

Growth primarily takes place near the soil surface at the plant's basal growth point. Initially, a new plant gets all of its energy from the seed, and seed-dependent growth is slow. New growth is also tender, thus making spring, with its more abundant moisture and lower temperatures, ideal for the new growth to put down roots before the sun really starts baking the land. Once you see sufficient green, leafy matter exposed above the basal growth point, the power plant (photosynthesis) kicks in and speeds up growth. As a plant reaches maturity, its growth slows, and the energy it creates through photosynthesis begins flower and seed-head production.

The objective is to manage pasture in order to maintain the plants' activity in the quick-growth phase. To accomplish this, you need to clip the plants just before their growth slows with flower and seed production, but you want to leave enough green surface showing to keep the power plant operating at full steam. Typically, this requires taking off about 40–50 percent of the leaf. You can do this mechanically, with a mowing device, or with those animated clipping units—also known as your animals.

Feed Definitions

- **Balanced ration**: A ration that provides all of the nutrients (including energy, fiber, protein, vitamins, and minerals) in the proper proportions for the animal's needs based on age and level of work.

- **Concentrate**: The grain or grains portion of the ration.

- **Dry matter**: The mass of the ration or feedstuff if the water is "baked off." For instance, a sample of mixed meadow hay might contain 85 percent dry matter, so your sixty-pound bale of hay would actually weigh fifty-one pounds on a dry matter basis (0.85 × 60 pounds).

- **Energy**: The part of the ration that is made up of sugars, fats and fatty acids, and starches used by the body for muscle and nerve activity, growth, fattening, and milk secretion.

- **Feedstuff**: Any food intended for livestock consumption.

- **Fiber**: The part of the ration that comes from cellulose and hemicellulose in plant matter and is broken down in ruminants and horses to create additional sugars and fatty acids.

- **Forage or roughage**: The hay or pasture portion of the ration.

- **Protein**: The portion of the ration that contains amino acids. The body requires protein for cell formation, development, and maintenance, especially for muscle and blood cells.

- **Ration**: The combination of foods in a specific diet, for a specific animal or class of animals, at any given time. It includes everything that the animal is receiving.

- **Supplements**: The vitamins, minerals, or protein added to the ration.

- **TDN**: Total digestible nutrients. These make up the portion of the ration that the animal can take advantage of. Feed reports, feed tags, and feed charts give the TDN of feedstuff. If the TDN on the previous sample of hay (see above definition of dry matter) was tested as 60 percent on a dry matter basis, the bale would contain 30.6 pounds of digestible nutrients (0.6 × 51 pounds).

After you graze (or mechanically clip) a plant, it requires a recovery period. The recovery period is the time it takes the plant to regain the energy it lost by being cut back—or the time that it takes for the plant to grow back to the length before you clipped it. If plants aren't allowed an adequate recovery period before they're bitten a second time, they weaken and may die, which can leave bare spots in your pasture or give rooms for weeds, those eternal opportunists, to move in.

Set-Stocking versus Rotation

Set-stocking is the most common grazing method. Farmers put animals into a pasture and keep them there indefinitely. Like kids in a candy store, the animals first go around and eat the things that they like best. Then, before their feed of choice has had an adequate recovery period, they come along and chew it down again. Meanwhile, the plants they don't like quite as much, or ones that have manure near them, never get chewed down at all, thereby reaching slow growth. The paradox is that both plants continue to lose energy, one because it is eaten too often and the other because it isn't munched on often enough.

Set-stocking results in overgrazing and overresting of plants in the same pasture at the same time. While the overrested plants do well in the short term, many of them are noxious weeds, so weed infestations usually increase with set-stocking.

You can manage grazing by subdividing the pasture into multiple paddocks, usually with electric

Better Grazing

You can improve grazing if you do the following:

- Aim for 40 percent legumes, such as alfalfa, clover, and birdsfoot trefoil, in your pasture.

- In a well-developed pasture, don't graze plants until they have reached at least 6 inches, and do graze plants before they reach 12 inches. In a newly seeded pasture, let the plants reach at least 8 inches before grazing.

- Remove 40–50 percent of the plant's leaf matter in each grazing period.

- During the growing season, don't graze plants lower than 3 inches from the soil surface. In winter, you can allow animals to graze plants to within about an inch of the soil surface.

- When the grass is growing fast during the spring and early summer, move the animals quickly; when it's growing slower, slow down their movement.

- If the grass is getting too far ahead of you (getting too long) during fast growth, mechanically clip it for hay or leave the clippings on the paddock as green manure.

fencing. Rotate the animals through before they have a chance to regraze the same plants, and keep the animals out of the paddock until the plants have had time to recharge their batteries. Generally speaking, the more paddocks available, the better. Paddocks may be either permanent or temporary. Four to eight permanent paddocks, which you can further subdivide with portable and temporary electric fencing, offer the most flexibility and allow you to time your animals' movement between paddocks.

The timing of animal movement from one paddock to the next is critical. In the spring and early summer, the grass is growing very quickly, and recovery may take only seven to ten days. Later in the summer, the plants may require a month or more.

As an example, let's say you've subdivided your five-acre pasture into five permanent paddocks. It's May, and the grass is growing quickly. When you move the animals out of a paddock, they can return in about ten days. With four paddocks left, you can move them every two to three days (ten days divided by four paddocks equals two and a half days per paddock).

Great Sources of Feed

Lawns, gardens, and orchards are great sources of feed. Use portable electric fencing to allow animals to graze the lawn. Don't leave manure on the grass too long because it will cause a dead spot and draw flies. You can also toss grass clippings, garden waste, and windfall apples over the fence. Just be careful that horses don't eat too many apples at a time because it could bring on an attack of colic. Pigs and birds, with their omnivorous diets, can eat many of the items we eat, so table scraps won't go to waste if thrown into the pig trough or poultry pens.

Now, consider the same setup, but in the long, hot "dog days" of July and August. The grass is growing much more slowly and requires about forty days to recover. If you move the animals out of a paddock, they'd have to spend about ten days in each of the next four paddocks (forty days divided by four paddocks). The problem during this period is that with ten days between moves, the animals are starting to bite the same plants twice. How do you allow an adequate recovery period with no regrazing? One solution is to use temporary subdivisions to cut the permanent paddocks in half. Now the animals graze each temporary paddock for only five days (forty days divided by eight paddocks), but the original paddock gets its full forty days to recover. As with other aspects of farm management, there are plenty of experts whose advice can help you reach optimal pasture management.

Poisonous Plants and Foods

It is tragic to lose an animal to food poisoning, but each year it happens to farmers around the country. Many toxins have no antidotes, so prevention is truly the best medicine.

The first step is to learn what poisonous plants grow in your area, something your county extension agent should be able to help you with. He or she will know what those plants are, what they look like, if their toxicity is strictly seasonal or continuous, and to which classes of stock the plants are toxic.

Some crop plants are poisonous at certain periods in their growth. For example, sudangrasses and sorghums can cause prussic acid poisoning if consumed while the plants are immature or immediately following a frost. Grazing of these plants at a stage where prussic acid levels are high can result in very quick death.

Animals can also become ill due to spoiled feed. Unfortunately, it isn't always easy to tell when feed is spoiled, but mold is the most common cause of spoilage. Cattle tend to be less sensitive to mold than horses, sheep, and pigs, but all species can develop health problems from mold.

Although mold itself generally isn't fatal, some kinds do produce fatal mycotoxins as by-products of their life cycles, as do some fungi. Mycotoxins are common throughout the world and can be found on both stored feeds and feedstuff still standing in the field. As mentioned, fescue poisoning, a common problem in many southern, Midwestern, and western states, is the result of an endophyte fungus that produces a mycotoxin. Again, talk to your county extension agent or your veterinarian to learn whether mycotoxins could be a problem for your operation.

Water

In the barnyard, as in the kitchen, water is the elixir of life. It provides structure to cells, transports and breaks down nutrients, flushes toxins from the body, and moderates the body's temperature. Just like a person, an animal can live without any food for quite a while. But keep it from an adequate supply of water on a very hot day, and it can succumb to heat stroke after only a few hours. Even during the winter, three waterless days can result in an animal's death.

One of the biggest challenges regarding water is keeping it clean. No animal wants to drink manure- or urine-contaminated water. But anyone who has watched

A pig enjoys a drink from an innovative waterer.

POISONOUS PLANTS*

Common Name	Scientific Name	Animals Most Affected	Season
Bitterweed	*Hymenoxys odorata*	Sheep, cattle	Spring
Water hemlock	*Cicuta* spp.	All	Spring
Larkspur	*Delphinium* spp.	All	Spring, fall
Pokeweed	*Phytolacca americana*	Pigs most affected; also cattle, sheep, horses, humans	Spring
Cocklebur	*Xanthium* spp.	Pigs most affected; all other animals	Spring, fall
False hellebore/Corn lily	*Veratrum* spp.	Cattle, sheep, fowl	Spring
Horse chestnut	*Aesculus* spp.	All grazing animals	Spring, summer
Oak	*Quercus* spp.	All grazing animals	Spring, summer
Mesquite	*Prosopis glandulosa*	Cattle and goats; sheep are resistant	Summer, fall
Yellow star thistle/Yellow knapweed	*Centaurea solstitialis*	Horses	Summer, fall
White snakeroot	*Eupatorium rugosum*	Sheep, cattle, horses	Summer, fall
Chokecherry/Cherry/Peach (leaves and bark)	*Prunus* spp.	All grazing animals	All seasons
Milkweed	*Asclepias* spp.	All	All seasons
Jimson weed/Thorn apple	*Datura stramonium*	All	All seasons

*This is by no means a complete list; in fact, it is estimated that there are approximately 300 plants in North America that are capable of causing poisoning in animals at some point in their growth cycle. Some, such as ryegrass, are common and widely used in pastures. In the case of ryegrass, the problem occurs in seed heads that become infected with bacteria or fungi.

Habitat	Effects	Toxicity
Flooded areas, overgrazed range	Vomiting, green nasal discharge, anorexia	Toxin is cumulative; avoid overgrazing
Open moist to wet areas	Salivation, muscular twitching, dilated pupils	Generally fatal, extremely toxic
Either cultivated or wild; the latter usually in open foothills and meadows among aspen or poplar stands	Arched back, falling, constipation, bloat, vomiting	Moderately toxic; causes death in some cases; young plants and seeds are most toxic
Disturbed areas with rich soil, pastures, waste areas	Vomiting, abdominal pain, bloody diarrhea	Mildly toxic in small doses but may result in death when consumed in large quantities
Fields, waste areas, edges of ponds and rivers	Anorexia, depression, vomiting, weakness, muscle spasms	Extremely toxic
Low, moist woods and pastures, mountain valleys	Vomiting, excessive salivation, cardiac arrhythmia, muscle weakness, paralysis, coma, birth defects in offspring of dams who have consumed it	Moderately toxic
Woods and thickets	Depression, lack of coordination, twitching, paralysis, death	Young shoots and seeds are highly toxic; otherwise, moderately toxic
Deciduous woods	Anorexia, rumen stasis, constipation followed by black diarrhea, dry muzzle	Moderately toxic when diet consists of more than 50 percent buds and leaves for an extended period
Dry ranges in brittle areas	Chronic wasting, excessive salivation, facial tremors	Moderately toxic when animals graze for extended periods; mixed-species grazing reduces losses
Waste areas and roadsides	Involuntary chewing movement, twitching of lips, inability to eat	Moderately toxic; death results from extended period of consumption, but horses eat it only in absence of other forage
Woods, cleared areas, waste areas, moist and rich soils	Weakness, trembling, weight loss, constipation	Extremely toxic, often causing death; toxins may be passed to humans through milk of affected animals
Waste areas, orchards, fence rows, dry slopes	Excitability leading to depression, lack of coordination, convulsions, bright pink mucous membranes	Extremely toxic; generally fatal in under 30 minutes, but animals who survive for two hours after consumption usually recover
Waste areas, roadsides, stream beds	Staggering, bloating, dilated pupils, rapid and weak pulse	Moderately toxic, but may result in death if sufficient amount is consumed
Fields, barn lots, trampled pastures, waste areas on rich bottom soils	Weak, rapid pulse; dilated pupils, coma	Extremely toxic; generally fatal within 48 hours

chickens trot through their water pan or a cow let loose next to a water tank knows that they are oblivious to the consequences and that clean drinking water for livestock is easier said than done! If I had a dollar for every time we had to drain and clean a tank with manure or urine in it, I'd be a heck of a lot richer. Automatic waterers can help reduce such occurrences, but even so, check livestock's water every day to make sure that it is clean and plentiful.

Winter presents its own challenges in zones where temperatures plunge below freezing. The days of chopping a hole in the ice of the pond behind the barn are thankfully behind us, but if you live where winters are frigid, you'll need to winterize your waterers to ensure that animals have sufficient hydration. There are a lot of options out there, ranging from applying heat tape to the pipes of existing automatic waterers to floating tank heaters or heated bowls for smaller animals. Neighboring farmers are usually glad to share their methods, as are livestock supply companies and representatives.

Shelter

Large animals that have their babies on pasture during the spring flush (the early part of the grass-growing season) and have shelter from wind and sun can get along pretty well without buildings. They can also use portable or temporary structures. Windbreaks and shade structures are easier to build and cost far less than a barn. We have raised all kinds of animals for decades and have done so without a barn more often than not. Don't get me wrong; a nice barn can be a great asset. But you don't need one to keep most large livestock, although you definitely need windbreaks and some available shade from the heat of summer. An open shed is ideal, providing relatively inexpensive wind protection, midday shade, and a dry place in inclement weather. If your critters are shielded from wind by walls or trees, you can even do without an open shed.

Rabbit hutches should be completely enclosed and raised off the ground.

Small animals, such as chickens, ducks, and rabbits, will need some type of structure for nighttime protection from both the elements and predators. It won't take predators long to see your farmstead as a place for free dinner when chickens are roosting unprotected or ducks are waddling around the yard near dusk. For poultry, a small traditional henhouse will work, or you can build a "chicken tractor," which is a portable enclosed shelter (some have wheels) that protects your chickens while allowing you to move them around your pastures for maximum productivity. Raise rabbits in individual hutches or in a pasture rabbit cage similar to the chicken tractor.

Alternatives to conventional poultry housing include straw bale structures, hoop-house structures, small metal or plastic hutches, and even tents or teepees. If you don't want to build your own, order them online or from a local builder.

Healthcare

Most animals are healthy if their basic needs are met. But, as with all living things, illness does happen. As mentioned earlier, sometimes animals get sick from poisonous substances, from improper diet, or from a sudden change in diet. However, most illnesses are generally caused by biological agents. Birds and mammals regularly act as hosts to a rather large menagerie of microorganic guests. Scientists estimate that a hundred billion microorganisms routinely share human bodies, and animals are no different. For the most part, this normal flora is harmless; under certain circumstances, however, these usually benign bugs can become pathogens.

When conditions are right for them, pathogens proliferate until their numbers simply overwhelm the animal, like weeds taking over a garden. Some produce toxins as a by-product of their bodily functions. *Clostridium* bacteria, for example, produce toxins that can cause tetanus, botulism, or black leg. Such toxin-producing pathogens are capable of causing illness with only a few organisms present. Other biological agents include viruses, yeast, fungi, worms, and other parasites.

We can treat bacterial diseases in animals with antibiotics, but, as we know from human medicine, not all bacteria respond the same way to all antibiotics. If a bacterium responds to treatment by a particular antibiotic, it's said to be sensitive to that antibiotic; if it doesn't respond to the treatment, it's resistant. When you treat an animal for a bacterial infection, it's best to have your veterinarian perform a culture and sensitivity test (unless you're dealing with an immediately life-threatening situation). This test will tell you which antibiotic will be most effective against the bacterium that's causing the illness.

As you probably already know, antibiotics do not affect viruses—not even a little bit. While some antiviral drugs have come on the market, for the most part, once a viral infection has begun, the immune system must produce antibodies to

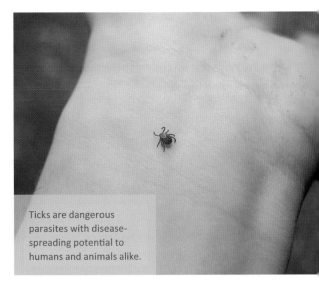

Ticks are dangerous parasites with disease-spreading potential to humans and animals alike.

The Power of Observation

Some illnesses and injuries cause readily apparent discomfort, but many don't. The power of observation is one of the best tools available to small-scale farmers when it comes to caring for their animals. So, what should you be looking for?

- **Listlessness.** Healthy animals are "bright-eyed and bushy-tailed," as the old saying goes. They are active, moving around freely. They hold their heads up and respond with their ears and eyes to their environment. They have a good appetite and drink plenty of water. An animal that is lying around, not eating, and not showing interest in its surroundings is probably ill. There are some exceptions to this rule. On very hot days, critters may just lie around in the shade, looking pretty lethargic, but as the heat of day breaks, they'll get up and eat again. Even if they're lethargic from the heat, they should still respond with their eyes and ears. Ruminants lie down to chew their cud, but no worries—as long as they're chewing their cud, they're feeling fine.

- Newborns can also be an exception to the listlessness rule. For the first week or two, a baby animal eats and sleeps, and its sleep tends to be very, very deep. Sometimes, you'll see a newborn that you think has died because it's in such a deep sleep, but when you go over to it, you realize that it's alive. Coming into the world from the safety and warmth of the womb is hard work for a little thing.

- **Sunken eyes or gray or white gums.** Sunken eyes usually indicate dehydration, which often accompanies illness. If the gums and tongue are gray or white instead of pink, chances are the animal is probably in shock—either from dehydration accompanying an illness or from an injury.

- **Poor-quality hair or wool.** The hair or wool should cover the body fairly evenly and look shiny and healthy (unless animals are shedding out their winter coats in spring and early summer). Poor-quality coats can indicate nutritional deficiencies, external or internal parasites, or other systemic diseases. In addition, if the tail head and the backs of the legs display caked-on manure, it is a sure sign of diarrhea or "scours."

- **Discharge from the nose, mouth, ears, or eyes.** While a bit of watery discharge from eyes or nose isn't anything to worry about, indicators that can mean an animal is not well are a discharge that looks like pus, crusty stuff built up around the muzzle or eyes, excessive slobber or frothiness around the mouth, or any kind of discharge from the ears.

- **Coughing or wheezing.** Healthy animals breathe easily through their noses, not their mouths, although sometimes they may mouth-breathe after excessive exercise or during extreme heat. Coughing is a sign that the animal may have an infection or that there's a physical irritant, such as dusty feed, present.

- **Hot, red, or hard udder.** Although most common in dairy cows, mastitis can occur in any female animal producing milk in its udder. On rare occasions, a young female that hasn't bred can develop mastitis. A healthy udder should be warm, not hot; pink, not red; and soft, not hard. The milk should flow smoothly and—except for colostrum—will be very liquid, with no clots or lumps. Colostrum is almost the consistency of pudding, but it shouldn't have any lumps in it after the first few squirts.

combat the virus, or the animal will die. However, although antibiotics won't cure viral infections, they may delay secondary bacterial infections, some of which can be deadly.

Bovine viral diarrhea (BVD) is a good example of a viral disease that is highly contagious and results in high mortality rates, meaning that large numbers of exposed animals will die. The virus itself isn't usually the cause of death; it's the secondary bacterial pneumonia that follows. So, as soon as a veterinarian diagnoses BVD, he or she begins the animal on antibiotic therapy. Veterinarians also administer drugs to help alleviate certain symptoms (for example, medications to reduce fever).

Viruses may not succumb to antibiotics, but you can prevent many common livestock viruses by having your animals vaccinated. Basically, a vaccine teaches the body to recognize the protein sequence of a given virus. Once the body's immune system has recognized the virus as a foreign invader, it will quickly recognize it again. This preprogramming allows the immune system to deploy antibodies when the virus first shows up, thus cutting down the immune system's response time to a point where the invading virus has little chance to begin reproducing. Your veterinarian can advise you on what vaccines your various animals need and how often; a large-animal veterinarian can travel to your farm to administer vaccinations.

Unlike bacteria and viruses, yeast and other fungi are multicelled organisms that normally don't cause problems in healthy animals. But when an animal's immune system is already compromised, multicelled organisms can cause a variety of skin problems, respiratory problems, and mastitis (infection of the milk-secreting gland of a female animal) of various types. Often, such problems occur following the use of antibiotics because the antibiotics upset the balance of normal flora, providing an opportunity for the invaders to multiply.

And then there are parasites. A parasite is an organism that lives in or on another organism, and it obtains food and/or shelter from its "host." Parasites aren't a single class of organisms. Instead, they run the gamut from protozoa (single-celled members of the animal kingdom) to far more complex organisms, such as worms and insects.

Some parasites are relatively benign. Others, such as biting flies, are nuisances; still others can cause serious illness. Tick-borne illnesses are on the rise, and West Nile virus, spread by mosquitos, is becoming a serious problem in some states. Parasites are capable of attacking most parts of the body. In cows alone, there are almost one hundred known pathogenic parasites. You'll find pathogenic parasites throughout the digestive system, on the skin, in the blood, throughout the respiratory system, in the eyes—everywhere.

Strong, healthy animals rotated on clean pastures are less likely to suffer from parasitic diseases. In the United States, intestinal worms tend to be the biggest parasite problem. Although not often fatal, intestinal worms reduce weight gain and milk production, and they simply tax an animal's system. There are medications available for treating worms; however, before treating, have your vet run a stool sample from one or two animals in the herd to check for worm eggs, which will be present if the worm problem is significant. If the vet doesn't find any eggs, then you don't need to treat the

Restraint

Throughout all of the health discussions in this book, I talk about what you should do and steps you should take, but if you can't control an animal, taking care of its problems is quite difficult. We work to get all breeding animals tame enough so that we can move them easily into controlled areas. From there, we can then move them into smaller pens, where restraint comes more easily. This is something to begin early. At times, you can exert enough control in a confined area, such as a stall, to do what you need to do without additional restraints, but often you'll need more control. Halters, ropes, nose rings, and twitches are all good tools for controlling animals.

If you are dealing with fairly wild animals, you'll need extensive handling systems that include catch pens, chutes, and head gates. These can be expensive, but plan on spending the money if you don't have the time, the patience, or the type of animals that can be tamed.

The risk of bloat is increased with leguminous crops such as clover.

entire herd. Parasite resistance has been found to be a genetic trait in sheep, so if you're planning on acquiring a flock, you should check into which breeds carry this characteristic.

Common Illnesses

Just like us, our animals are subject to a variety of illnesses. We get the common cold; so do they. We sometimes suffer from upset stomachs or diarrhea; so do they. What follows are some of the more common afflictions from which your animals might suffer.

Bloat

Bloat is limited to ruminants. When the rumen traps excessive quantities of gas, bloat occurs. In extreme cases, bloat can be deadly within an hour or two. An animal can have a hereditary tendency to bloat, but bloat is caused more often by other factors, including certain proteins in forage, rumen microbe efficiency, the quality and/or rate at which roughage is consumed, reactions to parasite treatment, or even choking. That said, bloat most often results from the animal's eating lush, leguminous pasture and is aggravated by moisture from dew or rain. It happens most commonly with alfalfa and clover. Pastures with a high percentage of grass compared to legumes are the least likely to cause bloat, but even these can do it in the early spring. In addition, if your cow gets into the feed bin and consumes too much grain, he can bloat as well.

Bloating can also happen in a cow that has rolled into a ditch or another low place and has trouble getting out. A cow stranded on its back continues to produce gas but cannot belch it out. Therefore, within mere hours, the buildup of gas can suffocate the animal because its lungs cannot expand. One Missouri farmer's milk cow laid down in a corner of the corral with her feet uphill, and if it hadn't been for a routine check and some helpful (and strong) neighbors, who literally lifted old Rosie to her feet, she would have been past help in another thirty minutes. Bloat can happen fast, so keep your eyes open and check on any animal that hasn't moved in a reasonable amount of time.

Simply put, when bloating occurs, foam begins to form in the rumen, hindering eructation (belching). The most prominent symptom of bloat is a bulge on the animal's left side, just below the spine and in front of the hipbone. This area normally appears caved in, but it protrudes in a bloating animal. Bloating animals also quit eating in addition to being unable to belch.

For bloating cows, various drenches, commercial or homemade, can help. There's also an old folk remedy of putting a rope through the animal's mouth, in a sort of "bit," and tying it around the back of its head; the rope causes the animal to chew continuously, releasing saliva, which will facilitate belching.

For a bloating cow, we mix together 1 cup of cooking oil, 1 cup of water, and 3 tablespoons of baking soda (use a quarter of this amount for a bloating sheep or goat, which is much less common). Using a squirt-top water bottle (like the kind that bikers and hikers use), we dribble the mixture into the animal's mouth for a few minutes; the animal won't consume all of the mixture but will ingest enough to be effective. After administering the mixture, we hold a smooth stick in the cow's mouth like a bit, which gets the tongue working, thus producing saliva and kick-starting the belching process. Once the animal starts belching, you can remove the stick, and you'll see the lump on its side start to subside.

In a life-or-death situation, it may be necessary to cut through the animal's side into its rumen to release the gas. Vets carry a two-part tool, called a trocar and cannula, for this purpose. In lieu of the trocar and cannula, a thin, sterilized knife (boiled in water or soaked in bleach for about five minutes) might save the animal's life. In either case, place the animal on antibiotics following the procedure, because infection is likely to follow.

Hoof Disorders

Laminitis and *grass founder* are related conditions of the hoof that can be risks for horses, especially certain breeds and many ponies. Laminitis is caused by the disruption of the blood flow to the laminae, the connective tissue that holds the hoof wall to the main bone in the foot, the coffin bone. In cases where the animal eats grass or legumes with a high sugar content, the laminae will become inflamed, thus blocking the blood supply, causing pain and causing the coffin bone to separate from the hoof wall. The coffin bone can start to rotate downward, eventually puncturing the sole of the foot; at this point, the horse has *foundered*.

Although treatment is possible in most cases, some animals experience complications and unfortunately have to be euthanized. The trend toward natural hoof care—trimming according to balance and for optimum function of the hoof mechanism (expansion and contraction)—has seen chronic founder cases restored to health when dietary issues are also addressed, although these animals may need to wear hoofboots on rocky terrain.

Because foundering can be serious, it becomes doubly important to monitor grass intake, especially if your horse is what is referred to as "an easy keeper,"

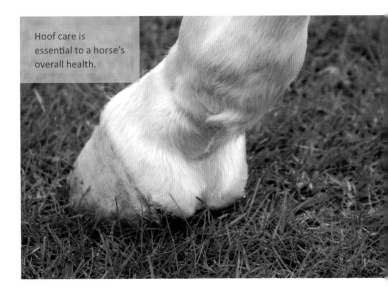
Hoof care is essential to a horse's overall health.

meaning that he puts on weight easily. One hobby farmer in Missouri owned a Morgan/Welsh Cob cross—a splendid calm mare whose breeding predisposed her to founder. After a bout that did not develop into anything serious, her owner was faced with the necessity of limiting her access to grass. Keeping such a horse in a corral or dry lot (small paddock with no graze) presented its own problems, so the owner purchased a grazing muzzle. This muzzle is essentially a basket with a small hole in the bottom, allowing the horse limited access to grass but free access to water. Wearing the muzzle enables the horse to be out in the pasture rather than alone in a stall or corral.

Hardware Disease

Unless you purchase a piece of completely bare land that has never had any buildings on it, chances are good that you will, at some time, have a run-in with hardware disease if you have any cows. Even baby calves can suffer from it. Although it's most prominent in cattle, it can happen in other ruminants as well.

When an animal eats a sharp piece of metal, such as a nail or a small hunk of wire, hardware disease can occur. The reticulum traps the piece of metal, which can puncture the wall of the reticulum or even the heart cavity. Symptoms include the animal's being in obvious pain, kicking at its side, getting up and lying down repeatedly, grunting, and having a slightly raised temperature. Inserting a magnet (specially made for this purpose) into the cow's stomach with a pill gun will attract the piece of metal and sometimes cause it to dislodge from the reticulum's wall; however, magnets are typically recommended as a preventative measure rather than as a cure. Some cattlemen insert magnets as a matter of course in all of their animals. Once inserted, the magnet remains in the reticulum for the life of the animal, attracting subsequent hardware that the animal might swallow, thus reducing the risk of puncture. If the problem is indeed hardware disease, the only way to remove the metal object is through surgery. Antibiotics are also administered to ward off any infection of the internal injury site.

Pneumonia and Other Respiratory Disorders

Respiratory illnesses can occur in all species. Stress caused by poor management (drafts or ammonia fumes in buildings, poor nutrition, and other hazards) or transportation of animals is often the underlying cause of respiratory illnesses because a stressed animal's normal flora can get out of control. Young animals of all species are susceptible.

If a mature animal has no fever and is still eating well, simply keep an eye on it. In young animals, a respiratory problem is, again, far more serious. Keep the animal warm, administer electrolytes, and call the vet if the condition persists for more than twenty-four hours or seems to be getting worse.

If your animals will come in contact with other animals (for example, if you plan to show them) or if you'll be bringing new animals into your herd, vaccination can prevent or greatly reduce the impact of many contagious respiratory diseases.

Scours (Diarrhea)

In adult animals, diarrhea, or scours, usually isn't fatal. Most often, adult diarrhea is the result of a change in diet or the consumption of very lush pasture. Mild cases from changing diets will clear up in two to three days and don't have many symptoms other than the diarrhea itself. Lush-pasture cases will continue for as long as the high-quality, moist feed lasts; however, like change-of-diet cases, it doesn't tend to have other symptoms associated with it.

If you haven't adjusted the animal's diet and it's not on lush pasture, the next most common cause of adult scours is excessive parasite loads. Animals don't typically have a fever with parasitic scours, but the animal will appear lethargic, and its coat may appear dull. Diagnosis of parasitic scours requires your vet to check a stool sample (unless you actually see worms in the stool). If an adult animal is suffering from both diarrhea and a fever, it's time to call your veterinarian. The animal is suffering from either a viral or bacterial case of scours.

Scours in baby animals is always a very serious and life-threatening situation. Normal baby-animal stools are yellowish, with a consistency similar to soft peanut butter. Sometimes the stools stick to the tail for the first day or two. During the fly season, wipe away stools to prevent screw flies from laying their eggs (which develop into maggots) there.

With scours in very young animals, the stool becomes watery or slimy and, if left untreated, the baby will die within a few days. Scours is quite common in bottle-fed babies, and the most prevalent

Electrolytes

Chemically speaking, an electrolyte is a solution—such as salt water—that will conduct electricity. When you dissolve things such as sodium, potassium, or chloride in water, they create an electrolyte solution. (Pure water, found only in a laboratory, won't conduct electricity.) Living creatures depend on a proper balance of electrolytes in their bloodstream. When illness upsets that balance, bodily functions (such as liver and kidney function) suffer. Most weak or sick animals benefit from the use of an electrolyte solution, although weak babies—especially those suffering from dehydration associated with fever or scours—must be given additional electrolytes if you hope to save them. Although sugar dissolved in pure water isn't electrolytic, people usually add it to electrolyte solutions to supply extra energy.

You can purchase electrolytes from farm supply stores or veterinarians to keep on hand, or you can prepare a homemade electrolyte solution. The homemade solution doesn't have the variety of ions and doesn't contain some of the added vitamins or probiotics that commercial products do, but when you need something in a hurry and the nearest farm store is an hour away, the homemade version can mean the difference between life and death. In 1 gallon of warm water, dissolve 4 tablespoons of corn syrup, 2 teaspoons of table salt, and 2 teaspoons of baking soda. (A friend of ours swears by the addition of a teabag to her homemade solution, and another friend uses a sports drink instead of the homemade solution.)

If you're treating a baby animal, dilute its normal milk ration by half with water. In between the milk feedings, feed it a comparable ration of your electrolyte solution. Don't feed the electrolyte solution with the milk because the digestive process interferes with absorption of the electrolytes into the animal's system.

cause is overfeeding, especially overfeeding of milk. Scours caused by overfeeding is the easiest type to treat, but without treatment, an otherwise healthy baby can die in just a few days. Baby animals are also subject to diarrhea caused by bacteria, viruses, and parasites.

As soon as you recognize a case of scours, the first thing to do is to replace fluids and electrolytes. Babies suffering from scours due to overfeeding often require no further treatment than this, but continue treatment for two to four days, or until the stool returns to normal, at which point you can gradually reintroduce milk to let the digestive tract adjust. If you do suspect that a pathogen caused the scours, antibiotics may be in order. Check with your vet.

Reproductive Issues

For most of us, one of the greatest joys of raising livestock comes from the miracle of reproduction. The opportunity to witness the birth of a calf, to see the first bumbling steps of a colt, or to spy a chick peeking out from under its mother's ruffled feathers is part of what draws us to animal agriculture in the first place. However, the reproductive process can also be a source of trouble.

When females release eggs from the ovaries (or from one fully developed ovary, as is the case in poultry), we call it the *estrus cycle*. Length of cycle varies from species to species in animals; in most, the female will allow the male to breed her only when the egg is ready to be fertilized, referred to on the farm as "in heat." Normally, cows and mares release one egg per cycle, so twins are unusual, and triplets are a real rarity. Ewes typically drop between one and three eggs per cycle, although some breeds are capable of releasing up to six. Sows are the big producers, releasing as many as twenty eggs per cycle. Multiple births are most often the result of dropping multiple eggs.

Single births are the most common type in cows, which typically release only one egg per cycle.

Occasionally, however, one fertilized egg splits in two at the beginning of development, the same anomaly that causes identical twins in humans. Chickens can lay about twenty-eight eggs per month, but eggs with double yolks—what we might call twins—do not usually develop two viable chicks.

The estrus cycle in some species, such as horses, sheep, and chickens, is seasonally cyclic—that is, the number of hours of daylight controls their cycles, and during part of the year, they do not come into heat at all (or their heat isn't very powerful). Cows and pigs cycle regularly all year long.

Fertilization may take place either naturally—through copulation between a male and a female—or through artificial insemination (AI). AI is now a common practice for most species of livestock. Dairy cows and turkeys are almost all bred artificially in commercial agriculture. The use of AI is also increasing dramatically in the pork industry as well. Farmers who utilize AI do not need to keep male breeding stock on site, yet they have access to high-quality sires, and the price isn't exorbitant. It's a good way to introduce certain genetics into a herd or flock.

As noted earlier, different species gestate for different periods—horses take eleven month, whereas sheep take a little less than six. Learn about breed-specific signs so you know what to watch for. In cattle, during the last few days before delivery, the cow begins to show signs

The mother/baby bond is particularly strong in cows.

that the big event is near. Her udder starts to swell with milk, and the area around the tail head begins to flatten out and appear sunken. Mucus begins discharging from the vulva, and the vulva itself looks puffy. Often, during these last few days, mothers may not act like they usually do. They often go off by themselves, avoiding people and other members of the herd; they may "talk" more than normal; or they may act quite restless, lying down and getting back up frequently. Close to the time of birth, a sac of fluid may show up and break.

Labor can be short, or it can last for hours. First-time mothers generally take longer for labor than those that have delivered babies in the past. For larger animals, the most common delivery position is front feet first, followed by the nose. After the shoulders pass through the cervix, the baby just about shoots out, but getting to that point can take time. Rear presentations, or *breech births*, are far less common than frontal deliveries, but they can happen, and it is not unknown for a calf or foal to get one leg or even its head turned back, which must be remedied if the baby is to deliver without endangering itself and the mother.

Troubled deliveries are really few and far between in the overall scope of things, but certain factors do come into play. One aspect to be aware of when choosing sires is size. A small Jersey or Angus cow has only so much room in her pelvis. Crossing a small breed with a big breed, such as a Charolais, Limousin, or Belgian Blue, can produce a calf that is too big for the mother to comfortably birth.

Over the years, we've had to assist in deliveries only a few times. As a rule of thumb, any animal that has struggled in labor for more than an hour without making visible progress or that is obviously weak and tired needs assistance. Do keep your veterinarian's number close at hand just in case. Sometimes it's hard to decide whether to make that call; farm visits aren't cheap. But if you've never run into a delivery problem before, go ahead and call a vet. In the best-case scenario, you don't need him or her, and your mother and baby are just fine. In the worst case, you wait too long, and, when the vet arrives, he or she can't save one or both animals.

After you've worked through the procedures with an experienced guide, you'll have a better feel for the situation and know what to do in most cases, but that first exposure definitely benefits from

Nursing usually comes naturally to the youngster.

guidance. Be cautious about asking other farmers for help and advice in this particular department; some may perhaps be a bit quick to pull babies, use too much force, or use the wrong kind of force, and this can ultimately cause problems rather than prevent them.

For mammals, once the infant comes into the world, the mother will lick down the mucus-covered newborn. Sometimes, a first-time mother is confused about what's going on. In this case, make sure to clean the baby's air passage (the area around its nose and mouth) so it can breathe. If the animal isn't breathing, tickle its nostrils with hay or rub vigorously with a gunny sack or some other rough material to stimulate breathing. Older methods included holding the baby upside down to drain excess mucous from the lungs, but this practice has been dismissed as counterproductive by most vets.

Baby animals need to begin nursing soon after birth for two reasons: one, to get the colostrum, and two, for the energy. According to the Colorado State University Extension, a calf should stand and nurse within two hours of birth. Watch from a distance, and if Junior hasn't found the teat by then, it may be time to get involved. This is easy if the mother is used to being handled. First, hand-strip a little milk out of each teat to make sure they are open. Then, nudge the baby into position, squirting colostrum into its mouth to encourage suckling, and slip the teat into its mouth. Some babies latch right on at this point, but keep working with it if it does not. Once or twice over the years in the case of especially weak babies, we simply had to strip a little milk into a bottle with a nipple and feed the first serving that way. This usually gave the animals enough energy to feed themselves.

A big beware: Mama bears are not the only mothers who fiercely protect their young. Don't expect even your family milk cow to invite intrusion. It is much safer to herd the new mom into a pen where you can tie her head and even hobble her feet if necessary so you can escape injury while you're getting the baby latched on.

Another aspect of reproductive issues is infertility, which can occur in both males and females but is more common in females. The main cause of infertility in females is an infectious process in the uterus following the birth of a baby, which interferes with her ability to have more babies. Some causes of such an infection can be a birth that required human assistance or the animal's failure to pass the afterbirth. If you haven't seen the mother pass the afterbirth, keep an eye on her for any discharge over the next week or two. Clear to slightly turbid, mucous discharges with some blood are quite normal, but if the discharge is thick and pus-like, have the vet come out and check the animal.

When male infertility does occur, it is a much bigger concern than female infertility because one male is responsible for breeding many females. If a female is infertile, she won't have any lambs, but the rest of the flock will still be producing. If the prize ram for which you paid dearly is a dud, there will be no lambs next year. When purchasing a breeding male, ask a veterinarian to check fertility levels/infertility problems.

A Closer Look at Chickens

In many ways, the chicken is the quintessential farm animal. From the iconic rooster on the ridgeline of a barn roof to eggs as brown as chocolate, nothing says "farm" in quite the same way. Chickens are arguably one of the easiest animals to acquire and care for, they produce food on a daily basis, and they can bring an abundance of pleasure as well. No wonder they can be found in farmyards, backyards, and urban henhouses from Portland, Oregon, to Portland, Maine, and everywhere in between.

A Bit of History

Chickens go way back in history. How far, no one can quite prove. What is known and confirmed by DNA testing is that chickens have in their lineage *Gallus gallus,* the red jungle fowl whose home range extends from the Philippines to northeastern India. Males of this wild species have the comb, red wattles, spurs, and the cock-a-doodle-do we're familiar with, while the females brood eggs and cluck like our beloved barnyard hens. But *G. gallus* is not the only contributor to the genetics of our modern-day chickens. Scientists have found three other species that they suspect of having mingled with the red jungle fowl, but the gene trail is incomplete, so our chickens' origins are still a bit of a mystery.

What is known is that, in 2004, a team of international geneticists mapped the chicken genome, and the chicken became the first bird and first domesticated animal so honored. Michael Zody,

Many hobby farmers enjoy raising chickens for the farm-fresh eggs they provide.

a computational biologist who studies genetics at the Broad Institute of Harvard and MIT, and a team of colleagues have been using the genome map to research the differences between the red jungle fowl and our chickens of today. The project, led by Uppsala University in Sweden, has found mutations that contribute to certain characteristics, such as year-round egg laying or significant weight gain.

Although evidence is not conclusive, it does seem to indicate that the Indus Valley was the area from which chickens spread westward, and we have cuneiform writing dating from 2,000 BC that mentions what is believed to be chickens. Some 500 or so years later, chickens showed up in Egypt, adorning tombs and hieroglyphs, but it took 1,000 additional years, more or less, before everyday Egyptians had their own backyard chickens. But once they did, they went into large-scale incubation, building ovens with vents to hatch multitudes of chickens and keeping their methods secret from outsiders for hundreds of years. Archeological digs in Mesopotamia have unearthed chicken bones circa 800 BC, and the Romans had chickens, too. In fact, the Romans liked them so much that officials issued a law in 161 BC that limited consumption of chicken to one per meal (not per individual) in an effort to curb moral and social decay. It sounds as if they were gorging on chicken way too often!

With the fall of Rome, chickens receded in importance, and, throughout medieval times, partridges and geese were the poultry of choice. In the Americas, chickens had an early start—some archaeologists believe that Polynesians introduced chickens to the Pacific coast of South America over a century before Columbus discovered the New World, but they were slow to take hold.

In North America, chickens were relative latecomers. However, because they were inexpensive and relatively easy to acquire, early settlers looked on chickens as more of a family hobby than as livestock because chickens required far less space and housing than the larger farm animals. Not only could the birds exist on scraps, bugs, and grain dropped by bigger stock, but they weren't a lot of work, either. Early American farmers typically let their chickens range, allowing them to roost in trees or in the barn and survive predators as best they could (not recommended). Chicken coops were not all that common.

Interest in chickens really took off in America after the showy Cochin breed arrived on the scene in the 1800s.

In the 1800s, the US chicken scene was charmed by the advent of the Cochin, a breed of chicken imported from China. Fat, fluffy, and friendly, Cochins were beautiful birds and exceptional mothers. It wasn't long until everyone and their brother wanted Cochins in their backyards; however, they weren't adapted to the scrappy life of the pioneer chicken. Cochins are too heavy-bodied to fly into trees for the night, and their feathered legs are a problem in muddy environments. Also, because they were imported and therefore more expensive, owners began protecting their investments by building chicken houses.

Cochins were so showy that, before long, chicken exhibitions became popular, and owners and farmers started selective breeding programs to produce more unusual birds to show off. Bit by bit, standard breeds—those that would have the same traits as their parents—developed, and so the hobby went until the latter part of the 1800s.

At this point, although valued for their eggs and an occasional Sunday dinner, chickens were not regarded as a major meat source in the United States. Even after hog and cattle consumption rose with the advent of mechanized slaughterhouses and huge feedlots, chickens remained a predominantly local enterprise. But all of that changed when medicated feed came on the scene, making it possible to house thousands of chickens indoors without the benefit of sunlight or movement—obviously not an advancement from the chickens' standpoint!

Chicken as a meat eventually became more popular than beef. It was inevitable that the overcrowding, prophylactic medications, and unhealthy growing conditions would render these large poultry operations questionable at best. Nevertheless, they continue to operate largely unabated because the meat is so much cheaper to produce than beef or pork. However, in a grassroots movement built on a desire to eradicate inhumane living conditions for the birds, coupled with the desire to eat healthy meat and eggs, people are reclaiming the realm of responsibly raised organic birds, and they're doing it one hobby farm and backyard at a time.

Backyard Chickens

As hobby farmers and backyard chicken enthusiasts will testify, the desire for healthy food and humane environments are only two benefits gained from keeping chickens. These birds are great assets in the garden as well. Not only can you use their composted manure to help fertilize your soil, but chickens are the elite bug patrol, searching out and snacking on every creepy-crawly within reach; as they scritch and scratch at the soil around your plants, they eat or obliterate small weeds as well. They also have a penchant for edibles such as strawberries and tomatoes, but, if they eat yours, be comforted by the fact that all of those nutrients are not lost but instead will show up in their eggs.

Chickens are popular on farms of all sizes as well as in suburban backyards.

Keeping chickens is a very "green" endeavor as well. If you're aiming for a sustainable lifestyle, this is a great place to start. Not only are you opting out of factory-farmed meat and eggs, these versatile birds are thrilled to eat any food scraps and leftovers that you wouldn't normally compost, which is much "greener" than running them through the garbage disposal or putting them in the trash can. In fact, chickens will also eat lawn and garden waste, and whatever they don't eat, they will scratch down into compostable material.

Many children enjoy participating in the care of chickens, which are much easier to keep than most livestock.

If you've been buying organic eggs, you know that they can really tax your grocery budget. "There's something so satisfying and self-sufficient about gathering eggs from your own hens, and the freshness can't be beat," says one enthusiast. "I love not having to run to the grocery store if I discover I'm out of eggs. All I have to do is pay a visit to the henhouse and gather their daily offering." However, a not-so-enthusiastic backyard chicken owner reported, "I had hopes of producing our own eggs for health, financial, and learning [benefits]. Reality? Here [in Montana], they aren't cost effective, especially during the winter, and [the chickens] cost more to feed than their eggs saved me."

If you're raising a family, caring for chickens can be a great way to help kids learn about the responsibility of caring for something that depends on them. Although many grow tired of feeding, watering, and the various other chores that go with keeping chickens, an informal poll by MyPetChicken.com showed that 8.3 percent of people who decide to have chickens do so because they had them growing up. That same poll registered that 35.6 percent of people wanted to keep chickens because they make awesome pets. Barbara Bartell, who you will hear more from in the chapter on bees, tells of two bantams she had while growing up: a blonde hen with feathers on her legs and a brown rooster. She would carry them around, and they, in their turn, followed her everywhere and slept in her doll's bed. Yes, chickens have their own personalities.

Other advantages:

- You don't have to clean their litter boxes.
- You don't need to take them for walks.
- They won't claw your furniture, leave hairballs on the rug, chew on your fence, or scare your neighbor's kids.
- They make great conversation-starters.

Getting Started

If you were vacillating before, but now you want to rush right out and get some feathered wonders, there are a few things to do first that will set you up for success. In their excellent book, *My Pet Chicken Handbook,* Lissa Lucas and Traci Torres discuss not only the desirable aspects of chickens but also the elements you'll want to think through before acquiring them. Among these are zoning laws (see sidebar on page 173) and whether you can commit to regular upkeep, including the daily feeding and watering, gathering of eggs, and periodic cleaning of the chicken coop. Not to discourage you—quite the contrary, I'm a chicken lover and veteran chicken owner myself! But I also know how "un-fun" it can get at times, especially if you're keeping chickens for their eggs rather than for their personalities.

When you must suit up and trudge through knee-deep snow to gather those eggs several times a day before they freeze (when you'd really like to be cuddling your cup of hot coffee or tea) or when you find yourself trying to locate someone who is willing to come and do your chicken chores because you'll be away, that's where the rubber meets the road. Hobby farmers sail right past most of these obstacles because chickens really are super-easy to care for compared to the bigger animals, but it's always good to know exactly what you're getting into before you launch.

Chicken-Keeping Options
Free Range

In this day of organics and sustainability, it's tempting to automatically choose to go free range with your flock. The basic method employed by most early settlers, *free range* means that your flock is roaming and foraging in a large yard or field with no fenced-in runs and no restrictions of any sort. Free ranging is less expensive because there's little initial investment if you don't build a coop (not recommended). You'll have reduced feed bills, especially in spring and summer, and you won't have a chicken coop to clean out. There should be few incidences of aggression within the flock because there will be no overcrowding, and your eggs will be even more nutritious due to all the fresh forage. What's not to love?

Well, for one, chickens have no concept of staying within certain boundaries. Breeds that can fly, will fly. They will try to cross roads, as the old joke goes, but many don't make it to the other side. In addition, chances are that if you have enough space to let your flock free range, you've also got predators on your perimeter, sizing up which hen to have for dinner. Yes, flying breeds will roost high enough to be out of the reach of coyotes, foxes, or the neighbor's bird dog, but some chicken-eaters hunt by day, and raccoons and other climbers won't be fazed by heights.

To deal with the bulk of the threats, most owners will shell out the money to build a coop in which to secure the flock every night and let their chickens out in the morning. This does necessitate you being on hand. Every night. Every morning. Because that one time you decide not to shut your flock in for the

Free-range chickens can wander and forage as they please.

night could be the night that you lose your favorite hen. Also, because they are free to scratch and peck wherever their little hearts desire, chickens won't hesitate to create dust wallows in your flower beds and poop on your porch. Yes. They do that kind of thing routinely and without malice. And all of those extra-healthy eggs? Sometimes they can be hard to find because chickens can find places to lay that you don't know about. Finally, true free-range chickens have a greater risk of exposure to illnesses/infestations from wild birds.

Confined Ranging

One step removed from ranging free, confined ranging is the most common way to manage chickens in a suburban or urban setting. Confined ranging consists of having a coop for the nighttime and allowing your birds to range in an enclosed area, such as a fenced run or backyard, so that they don't just wander anywhere. A large confined area has all the pros of free ranging with fewer cons.

However, if the area is too small, green will give way to bare earth as the birds eat and scratch away the vegetation, and their high-nitrogen droppings kill what they land on if not quickly removed. The chickens will still benefit from finding bugs, breathing fresh air, and soaking in sunshine, but if the greenery disappears, their eggs will have a bit less nutritional value (though they will still be fresh and organic if you're feeding organic feed).

Speaking of feed, you'll be shelling out slightly more money to feed your chickens because a confined area doesn't offer limitless places to forage. One way to offset these drawbacks is through the use of an ingenious concept that goes by the name of "chicken tractor." No, the chickens don't drive it. These tractors—which are actually mobile coops with small attached runs—can be moved every day or two so that your chickens are always on fresh forage. Some innovative chicken owners even have built chicken tractors that fit between their garden rows, allowing the birds to deal with

Pasture-raised chickens in a large enclosure with plenty of room to roam.

weeds and bugs, leaving fertilizer (uncomposted chicken manure should be used as fertilizer only *between* garden rows) and tilled soil behind.

Since there's a built-in coop on one end of the chicken tractor, you don't have to worry about being home in time to shut in the chickens for the night. Just be aware, though, that a chicken tractor will not keep predators out unless you cover the floor with wire, in which case you will cut down on your birds' ability

Chicken tractors offer a compromise between free-ranging and confinement that often works well for both chickens and owners.

to reach worms and take dust baths. If you don't put a floor on your tractor, your yard needs to be flat enough to allow the bottom of the coop to meet the ground firmly, without gaps, so that predators can't get in and chickens can't get out. Also, flighty, higher-energy birds don't do well in chicken tractors, so if you decide that this is the solution for you, look into more docile breeds.

Perhaps the greatest drawback to the tractor is that you will need to move it often. How much can you or your family lift? To help you decide on the chicken tractor that best fits your situation, visit a site such as BackyardChickens.com, which has many models of chicken tractors and coops built by chicken owners, some with plans and step-by-step DIY instructions that will inspire and empower you to build your own chicken dwelling.

Part-Time Ranging

This third way to manage your flock works well if you aren't able to lay out the money for a secure run that is large enough to avoid overgrazing or for a chicken tractor, which can be pricier than you may expect, depending on what materials you use for it. Instead, you can build a small run, secured from predators (see sidebar on page 179), so that your flock can access the outside every day, and then you can let your birds out of the coop or run onto the rest of your property to range only when it is convenient for you. People who use this method often let their hens out close to evening so that the chickens will head for their roost at sunset, which they will do every evening without needing your assistance. As a side note, as anyone who has tried this before will tell you, herding chickens toward their pen is an operation in futility, stressful for herder and chickens alike. Try it sometime. You'll see what I mean. They squawk and scatter, sort of like Chicken Little when she thought the sky was falling. But when it starts to get dark, they return to their coop like clockwork.

For some, part-time ranging must be supervised to head off predators or to keep your birds out of your garden, off the porch, or away from the road. An advantage of part-time ranging is that if you have a relatively small space in which the chickens can range, you can monitor how much time they spend outside the coop so that the ground does not become stripped of greenery. Part-time ranging affects the nutrient content in the eggs in a similar way as confined ranging. Also, you'll want to be sure that your coop and run give your birds sufficient space for the rest of the time, since crowded and/or bored chickens tend to pick on each other.

Full-Time Confinement

This option may sound terrible, but before you dismiss it, let's look at what it means. It does not mean imprisonment; rather, it means that a flock does not have any access to green pasture. The birds do, however, have access to the fresh air and sunshine of outdoors via a run, and this option does have some advantages over the others. For example, your chickens will be in their safe area all the time, and you don't have to be home to let them in and out or watch over them. Because their world is defined by their house and outdoor run, your gardens, flowers, and porch are safe. You will buy a little more feed because there will be no foraging, but not much more than what you would need in a part-time or confined-ranging setup; if your flock is small, the amount of additional feed needed is negligible.

Your chickens' eggs will reflect the lack of greens in their diet, but you can offset this by tossing all your of garden greens and extras into the chicken run. A confined setup may need a bit more maintenance—an additional cleaning or two over the course of a year—but, on the bright side, you'll have more material for your compost bin. Full-time confinement doesn't require as much space, either, because you need room only for the coop and run. Another plus is that the eggs will be in the nest boxes, right where they should be, not behind the lilac bush or deep in the woodpile.

An enclosed coop and run means less freedom, but more safety, for your birds.

Housing the Flock

Unless you opt for a free-range, survive-if-you-can style of keeping chickens, you'll almost definitely want a house for your flock. The good news is that you can build your own chicken coop relatively cheaply. Free plans abound (again, check out BackYardChickens.com). Some companies produce kits that you can assemble yourself, or you can purchase a fully assembled coop. You might consider hiring a local farmer or builder who should be able to throw up a chicken house pretty easily.

Whichever route you take, you can spend as much or as little as you want on a coop. Some suburban chickens live in Victorian coops with flowers in window boxes and vintage décor, whereas others are at the other end of the spectrum in an old doghouse set in a chain-link run with chicken wire stretched over the top to prevent predator access (the chickens are totally oblivious to grandeur or lack thereof). Here are the basics that you need to include in your setup:

Roof. This will shelter your birds from the elements. There are roof options ranging from lift-up to "green" roofs planted with herbs, flowers, or veggies.

Floor. Unless you're opting for a doghouse, floors are typically wooden, but some chicken owners use vinyl flooring in their coops to make cleaning and disinfecting easier. Cement also works; it's durable, it stands up to shovels, and it hoses down well.

Roosts. Roosts give the chickens a place to sleep. The North Carolina State University (NCSU) Extension Service recommends a minimum of 9–10 inches of roost space per bird, with approximately 14 inches between roost bars. Slant the roost ladder at a 20–40-degree angle to keep the chickens on the upper rung(s) from pooping on the birds on the lower rung(s).

Roosts should be made of wood with squared-off edges.

Nest boxes give the hens a comfortable place to lay their eggs.

According to Torres and Lucas, the best roosts are at least 2 inches wide, flat-sided, and made of wood (a 2 x 2 works well). Round roost bars cause the chickens to have to grip, which exposes their toes in cold weather, and metal roosts can contribute to frostbite in winter. Also, remember that chickens like high places to roost, so make sure that your nest boxes are lower than the roosts or the chickens will sleep—and mess—in or on top of their nest boxes.

Nest boxes. Include one nest for every four birds, plus an extra in case one of your hens decides to brood her eggs and won't leave the nest.

Windows. You should provide some, but not a lot, of light, as chickens tend to like laying in dim (not dark) environments and have less of a tendency to eat their eggs or fight with each other when the lights are low. Windows can be made of any material that lets light in.

Ventilation. If the window(s) can be opened and screened, that will work great to provide sufficient ventilation. If not, you'll need some form of built-in ventilation—a roof vent, a screened opening with a sliding cover, cracks in the siding, or something similar. Chickens are susceptible to respiratory ailments, and they need the clean air.

Waterer. A waterer doesn't have to be automatic, but it should be elevated to keep the chickens from washing their feet and pooping in it. You'll find a plethora of models (see sidebar on page 183) at a farm supply store or online retailer.

Feeder. The NCSU Extension recommends giving each bird a minimum of three linear feet of feeder. Not all feeders are created equal—some allow more waste than others (see sidebar on page 182).

Dust-bath and scratching areas. Some owners use litter boxes or kiddie pools filled with sand, diatomaceous earth, and/or wood ash.

Access door. Make sure that you give yourself easy access to the coop and your chickens.

Most coops also include a predator-proof outdoor run with a door from the chicken house into that run. And, while I'm on the subject of access, you might come across some compact coop designs that feature a lift-up roof instead of a walk-in door. This looks great at first glance, but picture yourself reaching in over the side of the wall to clean, feed, water, or gather eggs, and it may not be quite so cute, especially if you have to stand on a ladder to reach the floor during routine cleaning. This doesn't mean that you have to build a walk-in coop, but, honestly, if you can spend the extra money for a bigger coop and a regular door, you may find that extra size handy in the future. Just don't make it too big, or else the birds' body heat won't be able to keep the coop as warm in the

Your henhouse can be as fancy or as basic as you like, as long as it offers the essentials.

Varmint-Proof Runs

Building a predator-proof run isn't rocket science, but it does help if you think about your setup from a predator's perspective. If you were a fox or a raccoon, how would you get to those juicy birds on the other side of that barrier? Burrow under? Climb over? Push through? Raccoons will try to do all three, and they're very persistent. With their cunning little paws, they can lift or slide doors enough to squeeze through, but latches usually foil them. Hawks have been known to swoop down into an uncovered run, nab a chicken, and fly right back out. Coyotes and foxes will try to dig under the perimeter of your run and will exploit any hole they can discover. Possums, although usually more interested in eggs than meat, are also good at breaking and entering. Barn cats don't typically bother full-grown birds, but chicks or smaller pullets can be fair game.

One Missouri chicken owner reports that he routinely found a big old possum sleeping in a nest box if the kids forgot to shut the chickens in the night before, and a 5-foot-long black snake would appear from time to time and was once caught with an egg half in and half out of its mouth! (Note: Black snakes will keep the mouse population down in your barn, and an egg now and then is a small price to pay compared to the amount of feed that mice can and will eat.)

To secure the perimeter, NCSU Extension suggests burying the chicken wire 6 inches deep. Even more secure is the old farming practice of digging a 6–8-inch trench in which you place treated 2 × 4s end to end all around the outside of the run. Staple the chicken wire to these boards and, once secured, cover the boards up with the soil you previously removed. This additional step will prevent a predator from digging down and dislodging the wire enough to get in.

You'll also need a lid, or cover, on top of your chickens' area. This can be as simple as chicken wire fastened all around the top edge or a slanted shed roof. The latter has the advantage of keeping rain and snow to a minimum, which reduces the amount of mud in the run and makes for happier chickens when the weather gets bad.

winter as it otherwise would. The NCSU Extension states, "Allow a minimum of 2½ to 3½ square feet per bird inside the weather-tight coop, and an additional 4 to 5 square feet per bird in the fenced outside area." These are *minimum* requirements, but more space makes for happier chickens and can help prevent aggressive behavior.

Also, plan your coop with the consideration in mind that, in time, you may want to expand your flock. You can use extra space for keeping new birds separate during the introductory period, for brooding chicks, or as a recovery room for an injured bird (illness is another story; you'll need to remove a sick bird from the flock).

Few farmers and chicken owners regret wiring their coops for electricity. It's not necessary, but if there's a chance you'll be using a heated waterer, if you'd like a light when you're cleaning the nest boxes or checking your birds, or if you might opt for an automatic door down the road, it's easier to install electricity at the time of building than to retrofit.

As with every other building, location is key. Here are a few things you'll want to take into consideration when choosing the site for your chickens:

- How close do you want the chickens to your house? Remember—even without a rooster, chickens wake at the crack of dawn and go about their often-quite-noisy business. For the same reason, if you've got close neighbors, put the coop as far away as you can from their bedroom. On the other hand, if you're looking forward to watching your birds for enjoyment, can you see them from your favorite chair or porch swing?

- How much exercise do you want when feeding, watering, gathering eggs, and the like? In other words, how far are you willing to walk back and forth to the chickens' area, keeping in mind that you have to keep up with your routine tasks even in bad weather?
- Where is your water source? Stringing a hose from a faucet clear across creation or carrying full buckets can get old fast.
- Where is your compost bin or garden? How far do you want to haul the chickens' soiled bedding material?
- Where are your property lines? Will headlights, streetlights, or noise bother your birds at night?
- Is there good drainage for wet and wintry weather?
- If you are installing electricity, does your projected site have easy access to a power pole?

Feeding and Watering

Feeding chickens is pretty simple, but they do have certain nutritional requirements. If they're free ranging, they will get most of the required nutrients on their own through foraging, but, even so, you'll still need to feed them. The food's nutritional content becomes even more important with chickens that do not free range. Laying hens do well on a diet with a 15–18-percent protein content. If you're raising chicks for laying hens, you'll need to feed them chick starter until they're six to eight weeks old and have their feathers. This special feed is usually in "mash" form—finely milled little crumbles of mixed grain—with a protein content of 20–22 percent. Then you'll transition them to pullet grower, which contains 14–16 percent protein, from the age of six to eight weeks to twenty weeks. At this point, they are soon to begin laying eggs and should be transitioned to lay pellets/ crumbles (16–18 percent protein level). Typical all-purpose poultry food contains approximately 16 percent protein. While this is by no means a poor feed, chicks do better with higher protein to start out with, and layers are more consistent and healthier with a bit more protein than all-purpose feed gives them. If you plan on raising meat chickens, you will feed them a bit differently from egg-laying breeds: chick starter with 22–24 percent protein and grower at 16–20 percent protein.

Chicken feed contains corn and a mix of grains.

Once I switch to laying rations, I like to feed pellets rather than crumbles because when the hens bill their feed out of the feeder and onto the ground, pellets hold their shape longer than crumbles, so the chickens will find them again when they're scratching about in the litter. Crumbles sort of disappear into the dust.

The University of California Davis Extension states that after your laying pullets are fourteen weeks old or older, you can supplement ready-mixed feeds with grains such as corn, milo, wheat, or barley that might be cheaper, keeping the ratio to about ½ pound per ten pullets per day. This can be fed in a feeder or tossed onto the floor of the run or range; the latter affords them the fun and occupation of scratching and pecking it up.

Feeding too much grain can make your chickens fat, which is perhaps not a terrible thing going into winter, but, in reality, chickens don't need to be fat. Just healthy. Scattering whole or cracked grain (often called chicken scratch) on the floor of their run or coop helps keep the litter in good condition as the hens scratch and aerate it in search of grain. As I mentioned earlier, it also gives them something to do with their time, which makes them more content and less prone to picking on each other. Distractions help. If you decide to add whole grains to your flock's diet, offer them grit (sand) or tiny gravel two or three times a month so they can replenish the contents of their gizzard—the grain-grinding organ unique to birds.

No matter what age your birds are, they will love table scraps (not moldy or spoiled), grass clippings (as long as they don't contain pesticides/herbicides), garden surplus, and other supplements. All of these add variety and nutrients. Fresh or soured milk or other dairy products are also great. UC Davis Extension recommends putting milk products in plastic, glass, or enamel containers because the lactic acid content of dairy products can corrode galvanized metal, causing it to rust.

Chickens enjoy a mixture of greens and grains.

Chickens need an adequate amount of calcium to be able to form eggshells.

Laying hens need calcium. That's what eggshells are made of. If you're exclusively feeding a premixed laying ration with 2.5–3.5 percent calcium content, that is sufficient, according to UC Davis. But, if not, or if your birds forage or receive supplements of various types, make calcium available to them in the form of oyster shells or calcium grit (available through feed stores and online suppliers).

Some premixed feeds, especially chick starter, contain medications. Here is where organic thinking collides with typical recommendations, and you must forge your own way. Feed producers will, of course, advise and promote medicated feed as preventive against various poultry diseases. In reality, unless you're growing a lot of birds, your conditions are crowded, and/or your premises have a history of avian health issues, you don't really need medicated feed. That's sort of like giving your kids antibiotics on a prophylactic basis, just in case they might encounter germs. An added point to ponder is that any medications that your laying hens (or meat chickens) ingest will show

Feeders

There are two main types of feeder: trough and hanging (gravity-fed). Here are some pros and cons for each type as well as a few considerations.

Trough feeders

Usually made of plastic, metal, or wood.

Pros: They come in an array of sizes and styles, holding anywhere from 50 pounds of feed (a range feeder) to one for baby chicks that you can screw onto a Mason jar.

They work well for chicks and for free-choice feeding of oyster shell and grit.

Cons: They lead to food waste. Chickens often perch on and poop in trough feeders.

Quick fixes: Some trough feeders can be hung on a wall at the height of the chickens' backs or fitted with a grid or extra lip on the front side to help catch the feed as the chickens flip their bills sideways (as chickens do!).

If it is not a wall-hung model, you can elevate a trough feeder to the proper height by sitting it on cement blocks or a plastic platform.

Hanging feeders

Made of plastic or (usually galvanized) metal.

Pros: A hanging feeder gives the birds access to 360 degrees of feed.

Since it's up off the ground, chickens can't perch on the feeder, and they also tend not to bill out as much feed as long as the feeder is hung level with their backs.

Cons: Hanging feeders can be more difficult to fill with food than trough feeders.

A third type of feeder, which seems to have no drawbacks except the price, is a treadle feeder. It features a closed feed compartment, which is activated by the weight of the chicken and opens when the chicken steps on a platform in front of it; likewise, the compartment closes again when the bird steps off. This keeps the food clean and prevents feed waste from billing out as well as from rodent consumption. Grandpa's Feeders (www.grandpasfeeders.com) makes a sturdy and durable version.

up in their eggs or meat. Oregon State University Extension has this to say: "Residues in poultry are leftovers of compounds used in the production of birds to reduce or eliminate disease organisms or parasites. Coccidiostats, antimicrobials, wormers, mold inhibitors, and pesticides used for mites and/or lice control are potential residue producers. In addition, compounds used indirectly, such as rodenticides, insecticides, and herbicides, can become residues if birds inadvertently come into contact with them."

If medicating your chickens for a specific reason, be sure to follow the recommended withdrawal periods, keeping in mind that, in such cases, each day is a full 24 hours. The longer you can allow medications to clear from your birds' systems before harvesting their eggs or meat the better because you'll be the secondary recipient of any such residues. UC Davis recommends discontinuing medicated feed at sixteen weeks and/or at least five days before slaughter for meat chickens.

Can you mix your own feed? Absolutely, and more and more hobby farmers and chicken enthusiasts are doing it. However—and this is important—do your research to make sure that your birds' nutritional needs are being met. Just because your grandma threw some corn out every

Waterers

The best choices for waterers are pans, gravity-fed types, or automatic types. Here are some pros and cons for each type as well as a few considerations:

Water pans

Pros: Probably the only good thing about pan waterers is that you can rinse them and fill them easily.

Cons: Setting a pan of water out for chickens to drink out of is like inviting them to a community foot bath. Then, after they add a few droppings, they will finish by scratching debris into the pan.

Quick fix: Using a platform to lift a water pan up off the floor can help minimize the debris problem. However, birds like high places, and no bird worth its feed is deterred from perching on and pooping in an open water pan wherever it is located.

Gravity-fed waterers

Come in bottom-filled, top-filled, or two-piece construction of plastic or metal.

Pros: They come in a lot of different sizes and models.

They tend to stay cleaner than pans.

Plastic versions allow you to visually monitor the water levels.

Cons: Water can be heavy—5 gallons weigh approximately 42 pounds!

Gravity-fed waterers still gather debris and droppings.

Quick fix: As with feeders, an elevated waterer is going to stay cleaner than one that sits on the ground. If you have a ground-level model, raise the waterer to the level of your chickens' backs on cement blocks or some type of platform.

Automatic waterers

Types include a bucket that feeds into a cup or nipple and a nipple waterer fitted to a pipe that maintains constant pressure. Baby chicks will need their own waterers at their height with a trough small enough to preclude drowning.

Pros: The water stays free of litter and manure, giving your birds a constant clean water source.

Cons: Not many, other than the fact that biofilm (a mixture of fungi, bacteria, and viruses) can build up.

Quick fix: Scrub and disinfect the waterer periodically.

When choosing your feeder or waterer, consider practicality, cleanliness, space, and who ultimately does the feeding and watering, along with durability, ease of use and cleaning, accessibility, economy, and weather.

morning and her chickens survived does not mean that you can do the same unless you duplicate her circumstances—likely free-range birds that were getting most of what they needed through foraging and table scraps.

All self-mixed feed is not created equal. It is only as good as the ingredients that go into it. Sadly, with genetically modified organisms (GMOs) dominating corn crops and other grains, plus the pesticides and herbicides used routinely on those crops, mixing your own feed doesn't ensure that it's any healthier than premixed feeds unless you buy organic, non-GMO grains. Mixing your own organic feed also tends to be more expensive than buying organic feed because big companies can purchase organic, non-GMO grains for their mixes at a much lower price than you can. That said,

A trough-style chicken feeder should be hung level with the chickens' backs to prevent them from stepping or pooping in their feed.

No matter what type of waterer you use, be sure that your chickens have constant access to clean, fresh water.

some hobby farmers are taking sustainability to a higher level, growing their own organic grains to use as elements in their home-mixed feeds.

As far as the practical aspects of feeding chickens go, you will need different sized feeders as your birds grow. A good rule of thumb is to keep the trough at the same height as your chickens' backs. This will help prevent them from standing in or on the feeder, pooping in it, scratching litter from the floor into it, and scratching feed out onto the floor almost as fast as you can put it in. Avoid excessive waste by not filling troughs too full, and, for birds older than six weeks, have them clean up the previous day's food before feeding more. Allow 3–4 inches of feeder space per bird, and monitor how much food is left at the end of a day. This will help you figure out how much they need. If your feeder is an automatic one that holds more than one day's worth of feed, be sure to hang or position it to avoid excessive billing out and waste of feed. Chickens can handle eating on demand, and if you can reduce waste, this is the easiest way to go.

Watering is pretty straightforward—just make sure that your chickens have plenty of fresh, clean water at all times. Easily said but, with chickens, not always as easily done due to the fact that chickens don't think things through—things like "walking in my water pan will result in dirty water." No. They walk and poop with impunity, and waterers are not off limits for these activities. Just remember that chickens need about twice as much water during hot months as cold, they can become ill from contaminated water, and, in the more frigid parts of the country, you'll need to figure out some way to keep their water thawed.

Egg Production

Chickens begin laying eggs at around sixteen weeks old (some breeds may lay a bit earlier and some a bit later) and will continue to lay for five to ten years thereafter, peaking at about two years. The first eggs that a hen lays will be small, but they will gradually increase in size until they reach the size that is normal for that breed and that hen.

The process by which hens produce eggs is brilliant. A hen's reproductive system is made up of the ovary, in which the ova (we call them yolks) develop. Every twenty-four to twenty-six hours, the hen ovulates, releasing a mature ovum into the oviduct. This oviduct is a tube of approximately 25–27 inches long with five major sections—the infundibulum, magnum, isthmus, shell gland, and vagina.

If fertilization is going to take place, it happens in the fifteen to seventeen minutes during which the ovum is in the infundibulum. From there, it moves into the magnum, the longest section of the oviduct (13 inches). Over the course of approximately three hours, the yolk travels through this part of the oviduct while thick albumen (egg white) forms around the ovum. At the end of the magnum, we find the isthmus—a narrower section about 4 inches long in which the inner and outer shell membranes are formed over the course of about an hour and fifteen minutes, give or take a few. The fourth section—the shell gland or uterus—does exactly what its name implies. It is here that the egg acquires salt and water, and then calcification begins.

Over the next twenty or so hours, the hen uses 8–10 percent of her body's calcium to make

Egg Color

As if having your own fresh eggs is not fun enough in itself, eggs come in a range of colors from white or cream to deep, dark chocolate, and from pale sky-blue or green through to aqua, sage, teal, and other shades. Araucanas lay blue eggs. Thought to have originated in Chile, they are rumpless (without a tail head) and often have tufts of feathers hanging by fine, elastic skin threads to each side of the head. Sometimes birds will be marketed as Araucanas and will lay blue eggs, but if they have tails, they aren't Araucanas. They might, however, be Ameraucanas. This easygoing breed is more common.

What many hatcheries sell under the name "Araucana/Americana" (watch the spelling) are really "Easter Eggers." Easter Eggers are basically the mutts of the blue-egg-laying chicken world, but Lucas and Torres refer to them as "fabulous birds"—smart, friendly, hardy, and good layers. Their eggs range from blues and greens through to browns of all shades, reflecting their mixed breeding. As an added bonus, their eggs are large.

Cream Legbars also lay blue eggs and are great foragers, but they tend to be a little more flighty and less docile. Brown egg-layers abound, and some of the darkest eggs are laid by breeds such as the Black Copper Marans, Penedesenca, and Welsummer.

One fascinating aspect of egg color is that brown is applied to the surface of the white calcium-carbonate eggshell and thus can be washed or rubbed off when first laid. However, blue color is incorporated through the whole shell. So, if you have an Easter Egger laying olive eggs, what you've got is an egg of some shade of blue with an applied wash of brown.

Weird Eggs

We're used to seeing perfect eggs, so when something strange comes along, it has us asking questions. Here, then, are some facts:

- **A double-yolked egg** happens when the ovary matures and releases two ova at the same time. Young hens still regulating their systems tend to be the ones that lay double-yolked eggs. The egg itself is nutritionally the same as a single-yolked one. According to the Penn State Extension, it is rare to hatch two live chicks from a single egg because one embryo usually out-competes the other, which dies before hatching, or both of them don't make it.

- **Mottling on the yolk.** Lighter spots, usually unnoticeable on the paler yolks of commercial eggs, are not a problem and do not alter the nutritional content. Sometimes, excessive mottling can be a result of too little calcium in the hen's diet or the presence of cottonseed or sorghum.

- **Blood spots.** Usually found on the yolk, these are the result of a broken capillary at the time of ovulation. Incidents occur more often during periods of high activity in the hen. Don't throw the entire egg away; just discard the part with the spot.

- **Meat spots.** Brown in color and usually found in the egg white, they are tiny bits of the wall of the oviduct that slough off and get encapsulated with the ova as the albumin develops. They don't look particularly appealing, but you can still use the egg. Just scoop the spot out with a spoon and discard.

- **Shell-less egg.** Once in a while, the shell-making process doesn't take place, and you'll end up with an egg that has comes out without a shell. This isn't a problem unless it happens regularly, in which case you might suspect issues ranging from a deficiency of calcium, phosporus, and/or vitamin D to egg-drop syndrome or infectious bronchitis, all of which can cause increased incidence of shell-less eggs.

- **Yolkless egg.** Erroneously called a "pullet egg," a yolkless egg is actually a portion of the oviduct that has sloughed off, traveled down, and received shell before exiting. It actually looks a lot like cloudy egg white inside, but don't eat it! FYI, pullet eggs are the smaller eggs produced when a hen first begins to lay and are just regular eggs in miniature with a yolk, albumen, and shell. That said, yolkless eggs do tend to happen more with pullets than with mature layers.

- **Egg within an egg.** Rarer than a yolkless egg, an egg within an egg occurs when an egg that has already received a shell and is ready to be laid reverses for reasons yet a mystery, traveling back up the oviduct until it encounters the next ovum coming down. Albumen covers the whole shelled egg and new ovum together and then it moves on through the isthmus, receives the inner and outer membranes and shell, and is laid.

the shell, 47 percent of which is derived from her bones, and the remaining 53 percent from her food. Shells are laid down in the form of calcium carbonate, and, according to Purdue University, brown eggs acquire their color during the last five hours in the shell gland (see sidebar on page 185). The last section of the oviduct, the vagina, is 4–5 inches in length. The egg enters the vagina fully formed, small end first, and it turns around while in the vagina, ready to be pushed out into the world in just a few short minutes. Time for the "happy egg song"—*BOCK-bock-bock-bock-buck-OCK!*—to announce her achievement. The total time a hen's body takes to transform a yolk into a fully developed egg and lay that egg is about twenty-five to twenty-six hours. Whew! And then, about thirty to seventy-five minutes later, the whole process begins again.

Because the hen's egg cycle is regulated by the number of daylight hours, ovulation almost never occurs after 3:00 p.m. This means that if a hen lays an egg late in the day, her next ovulation will wait until after sunrise the next day, which is why, every so often, a hen will have a day when she does not lay an egg. Because of the egg cycle, you're not going to get more than one egg per hen per day, no matter what you feed her. This amounts to six or fewer eggs per week, per hen, depending on the breed as long as there are twelve to sixteen daylight hours in each twenty-four-hour period.

Hens will lay eggs whether or not a rooster is involved.

As the days shorten in the fall, chickens will molt (lose and replace feathers), and egg production will drop. Just how far it drops depends on the particular breed of hen—some will almost completely quit laying for about two months, whereas others will not experience quite as dramatic a drop. In order to prevent this seasonal drop in production, some owners will introduce artificial light in order to maintain twelve to sixteen hours of daylight. On the surface, this seems like a fine idea, but upon further consideration, it may not be all that wise. Chickens need to molt. If you delay their molting, there's a real possibility that they may start losing feathers when they need them the most—in the dead of winter. Obviously, that's going to work against the chickens' health, comfort, and egg production because they will use more energy to keep themselves warm, but without a full complement of feathers to help trap body heat, they may be chilled and at greater risk for frostbite and illness.

Weird Shells

Contrary to what you buy in the store, not all eggs are uniform. It's just that the strange forms don't make the grade. In your own coop, though, you may periodically encounter a number of aberrations.

- **Body check**. Sometimes an eggshell becomes damaged while it is being formed, and the shell gland repairs it. You can see the repair because it leaves a sort of "ring" as if progress was halted for a moment.

- **Thin spots and ridges**. These indicate weaker areas in the shell, although the egg inside is normal. Don't use these to incubate.

- **Strange shapes**. Irregular shapes include football (without a large end), pear, and other variations. Again, nothing is amiss inside these eggs, but they don't fit into cartons well and should not be incubated because their shapes may keep them from being rotated during incubation as they should be.

- **Warts and bumps**. These spots simply indicate extra calcium applied unevenly. Although they look weird, they are not a problem.

The Rooster's Role

Can a hen lay an egg if there is no rooster involved? Definitely. It just won't be fertile (able to hatch). Not a problem for 90+ percent of the chicken-owning population, since they're probably not wanting to raise chicks anyway. This is good news for suburban chicken owners because roosters can be too noisy for neighborhoods (and are usually banned by zoning laws). But roosters do more than fertilize eggs. Your flock will feel more secure with a mister or two on guard; they really do watch out for the hens, warning them when to run or take cover. In the pages of *My Pet Chicken Handbook,* Lissa Lucas recounts a time when her rooster, Gautier, gave his "predator-on-the-ground" warning cry. Lissa rushed out, barefoot and nightgown-clad, joining Gautier as he chased after a fox that had nabbed one of the hens and was dragging her down the hillside.

Rhode Island Red roosters are impressive in appearance while the hens are prolific layers.

Due to the rooster's watchful eye and brave defense, they were able to foil the fox and rescue Marissa, a Rhode Island Red/Faverolles hen, who ended up slightly befuddled but apparently none the worse for her ride downhill in the fox's mouth! She would have been a pile of feathers had it not been for Gautier (and Lissa, who cleared a barbed-wire fence on the fly).

A rooster is not only watchful and protective, he's forever on the search for tasty morsels. Upon finding them, he will issue a distinctive call that brings the girls running to devour whatever he found. He usually doesn't get much of the treat himself, yet he doesn't seem to mind. In addition to these practical aspects, roosters are a thing of beauty, usually much showier and of more colorful plumage than the hens. And, of course, roosters do make picturesque and musical alarm clocks, but these feathered reveille blowers sound off pretty much throughout the daylight hours. All of them. This can get old, especially if the coop is close to your house or your neighbor's.

A White Leghorn rooster. In addition to fertilized eggs, roosters bring other benefits to your flock.

If you keep one or more roosters, Traci Torres of MyPetChicken.com and *My Pet Chicken Handbook* recommends a ratio of one rooster to ten to twelve hens. With fewer hens, your hens may end up with broken feathers or bare necks or backs, which is neither pretty nor beneficial, especially in winter.

Choosing Your Birds

When deciding what breed of chicken best fits your world, it helps to know that chickens have body types just like people do. In chickens, these are called type A and type B. Type A birds put most of their nutritional intake into egg production, which makes them lean, tending-toward-nervous, often flighty egg-laying machines. Leghorns belong to this group. Type B birds lay eggs, too, but because they convert more of their diet into body weight, they are dual-purpose birds—meatier and producing fewer eggs per week than type As but with calm, easygoing temperaments. Dual-purpose birds are the kind that can go into your cooking pot after their egg production is declining, if you so desire.

Torres does the egg math for the rest of us. "If you want two dozen eggs a week or so, you'd only have to keep about four Rhode Island Reds to hit that target number. But if you want to keep funky, crested Polish hens, you might need twelve of those—or more—to get the same number of eggs. And even then, the Polish eggs would be smaller than the Rhode Island Red eggs, so you'd have to use more of them in your recipes."

And then there are chickens bred and raised solely for meat. There are not as many choices here. The University of Minnesota Extension states, "The most economical meat production is obtained from the commercial meat strains developed from breeds such as the Cornish, Plymouth Rock, and New Hampshire. These crosses have been bred for the most economical conversion of feed to poultry meat; they feather rapidly and mature early."

They're bred to mature quickly—seven to nine weeks for broilers/fryers (weighing about 2½–4 pounds dressed). If processed at five weeks of age, these birds are called Cornish game hens. When

Marans hens are known for producing dark chocolate-colored eggs.

If you're considering chicken breeds based on looks, you can't go wrong with the Polish.

grown longer—twelve weeks—they make delicious roasters. However, due to their breeding, the heavier they get, the less able they are to move around or access food and water, and, in some cases, their hearts give out. Raising them in chicken tractors to give high-quality forage, sunshine, and adequate space will lower their stress, and they will do quite well.

Chicks

Unless your hens are brooding and hatching their own chicks, if you want to start with chicks, you'll need to know how to keep them alive and thriving. The art of brooding chicks—raising them from the time of hatching up through the age at which they are fully feathered and able to live without extra warmth—has a bit of a learning curve, but it's generally not that difficult once you understand the particulars and get the right equipment. You will need:

Among the rewards of chicken keeping is the joy of raising your own chicks.

Predator-free, draft-free space. You'll need about 2 square feet per chick to brood them from hatch to six weeks old. This can be anything from a corner of the garage to a plywood or cardboard box. Just know that if you brood them in the garage or anywhere in the house, they will generate a *tremendous* amount of dust.

Brooder light. Newly hatched chicks must be kept between 90–95 degrees Fahrenheit to avoid hypothermia and death. Lucas and Torres recommend a 250-watt infrared heat lamp. Alternately, a Brinsea EcoGlow brooder (or similar product) is a small plastic tunnel that doesn't get as hot as a heat lamp, reducing the risk of fire but keeping chicks warm. Although more expensive than a heat lamp, it has the added advantage of not providing light around the clock.

A brooder light helps the hen keep her babies warm.

Chick feeder. Provide for 1 linear inch per chick, increasing to 2 linear inches at two weeks and up to 4 linear inches for growing pullets six weeks and older. Raising the feeder to the chicks' back level as they grow will help minimize waste.

Chick starter. Have chick starter on hand, and keep their feeder filled. Chick starter is milled finely to make it easier for chicks to eat. They'll be eating more and more as they grow, so you might need to add a second feeder in order to give them sufficient food.

Waterer. At first, you will need a narrow waterer that chicks can't climb into and drown. As they grow throughout their first six weeks, gradually raise the waterer to keep it at back height to maintain cleanliness. Unless your newly arrived chicks are showing signs of stress, they will do best with plain, rather than medicated, water or sugar water.

Bedding. As Torres says, "Baby chicks poop. A lot." Make sure that the floor of the brooder has at least 2 inches of absorbent bedding. She recommends pine shavings, but not cedar, as the oils may irritate the chicks' lungs. Straw is less absorbent, rots, molds easily, and is prone to insects. Hay has the same problems as straw but is even more prone to insects in addition to being more expensive. Torres also advises to avoid shredded paper. It's not very absorbent, and it can be slick, leading to splayed leg deformities. Plus, it can mold. Some people love using paper towels, but they're quite expensive and not all that easy to change.

A hen will keep her babies warm, but, in the absence of a hen, you'll be doing that job with a brooder light. Once you've ordered chicks, set up your brooder. Several days before their expected arrival date, turn on the heat source, monitoring and adjusting until it is just right so it will be ready if your chicks arrive early. They'll do best if you put them directly into their new warm home.

Start by hanging the light about 15 inches from the litter/bedding. Place some kind of shield around it to help direct the heat downward. The University of Minnesota Extension advises starting chicks at 90–95 degrees Fahrenheit (measured at 2 inches off the floor, under the edge of the hover) and then lowering the temperature "by 5 degrees per week until the supplemental heat is no longer needed. Observe the chicks to gauge their level of comfort."

You'll know that your heat level is right if the chicks ring around under the perimeter of the light. If they're crowding in the center, directly under the lamp, it is hanging too high and they're not warm enough. If they're ringed up outside of the circle of light, it's hanging too low and they're too hot. Chicks raised by a hen can come and go, warming up under her and then making little forays out into cooler air. Make sure that your chicks have enough room in the brooder to imitate this behavior.

Baby chicks really are adorable with their fluffy down and contented little cheeping noises. You'll notice that baby chicks sleep a lot, just like all newborns, but bit by bit they stay awake for longer at a time. Check them every few hours to make sure that they're staying warm, but not too warm. Watch their huddle, listen to their cheeping, and make adjustments if need be.

Chicks purchased as the spring days warm up will find the transition from brooder to coop easier than will chicks hatched in the fall that have to transition from the warmth of the brooder to life in the winter weather. However, even though the days may be warming up, spring chicks can become chilled during shipping. Torres recommends looking at the average temperatures for your area and scheduling your chicks' arrival at a time when your area's temperature will be above freezing and then will be at least 60 degrees Fahrenheit on the average six weeks later.

Ordering chicks is exciting! You can usually order pullets only, roosters only, or straight run, which is a random assortment of boys and girls. Pullets are higher priced, but you will come out with more layers that way, and you won't have to figure out what to do with all of the roosters if you can't keep them or don't want to eat them. Chicks raised for meat are usually shipped on a straight-run basis.

And then comes the big day when your chicks arrive! Count them, check your packing slip, and then inspect each chick for "pasting" as you transfer it to your brooder. Pasting happens when droppings dry around the chick's vent and seal it shut. Droppings on the down are not a problem, but a plugged vent is. Remove stubborn pasting with a warm wet paper towel, making sure that the chick doesn't get chilled. Keep an eye on any chick that had pasting to make sure that it doesn't occur again. On another note, some chicks will arrive with an umbilical cord still attached to the rear, a bit beneath the vent. Just leave it alone. Removing the scab or pulling the cord is painful and can kill the chick.

Let the chicks warm up as you watch them and adjust the heat as mentioned. When they're all warm and happy, encourage them to drink by dipping their beaks in water. Torres suggests putting a few sterilized marbles or stones in the waterer because chicks are curious about such objects and are likely to peck at them in the water, thus eventually learning how to drink. If you've got lovely weather—not too cold or breezy—give your chicks access to sun and shade after a few weeks in the brooder. Make an improvised "playpen" on the grass to give them access to fresh greens and bugs for short sessions. Their mother would do this. Watch that they don't become chilled, though, and that a cat doesn't hop in and decimate your flock.

You will need to keep the chick brooder clean and dry in order to minimize chances of illness. Also, sprinkle a bit of grit where the chicks can pick it up at will. Barring anything unexpected, your chicks should do just fine. At age six weeks or when the brooder light has been raised to the point where they are acclimated to the real world's temperature, you can transition them to their next home.

If you're introducing young birds into a grown flock, you're going to have issues—the youngsters will be seen as intruders and attacked. If you've got the space, keep the two flocks separated until your new chickens reach close to adult size so that they can survive the inevitable shuffle and scuffle as the updated pecking order emerges. Then introduce them all at once. Some chicken owners add an extra step by introducing the old hens into the young hens' space, which puts the old girls into new territory, making it easier for the young ones to find their way in the new pecking order. Either way, don't add just one or two new hens to an established flock. The established birds will make the newcomers miserable and may even kill them. Also, add a second feeder and waterer so that the low chickens on the totem pole can still eat and drink if they are chased and bullied away from one feeder and waterer.

Try to make the transition from brooder to coop in mild weather so that the young chickens don't get chilled or overheated. If these are fall chicks and you're headed into winter, help them adjust by improvising an interim coop that is a bit warmer than their ultimate destination. Alternately, you can hang the brooder light in the coop to help take the chill off for a bit. However, remember that chickens scratching about in the litter generate a whole lot of dust, which can become a fire hazard with heat lamps.

Baby chicks will enjoy life outside the coop in nice weather.

Don't heat your coop long-term. Chickens need to shift metabolism to acclimate to lower temperatures, but if you heat their house and then lose power during a temperature plunge, you could lose your flock. Even without that, chickens are like kids—it's hard to get them to go outside in the cold when their house is warm. Only heat during a precipitous drop in temperature while you're easing the chickens' transition. Furthermore, don't insulate your coop too well, because the moisture generated by the chickens' breathing and droppings can't evaporate, but may condense and freeze if the temperature drops, thus contributing to the risk of frostbite and respiratory and mold-related illnesses. Too much or not enough insulation can contribute to ammonia build-up, which damages chickens' lungs.

Other things to remember in cold weather:
- Give your chickens access to the outside. They'll utilize it.
- Gather eggs more often to keep them from freezing.
- Make sure that the chickens' water is free of ice.
- Keep the coop clean, even if it's cold out there.
- Protect their combs. If you live in a very cold area and have birds with large combs, apply a thin layer of petroleum jelly to keep the exposed skin healthy and free from chapping.
- Don't let your birds get bored because they could get aggressive. Torres suggests adding treats, such as hanging a head of cabbage for them to peck at or giving scratch grain to keep them occupied.

Give your birds a few weeks to settle down and own their new home. Once they do so and learn where it is, they'll be able to find it if you let them out to range.

As you can see, keeping chickens can be a fascinating, enjoyable venture, whether you do it in your backyard or on your farm. Just be warned—once you get bitten by the chicken-keeping bug, you may be affected for life!

Beekeeping

Ancient. Wise. Indispensable not only for honey's golden sweetness but also for pollinating a third of all of the food we eat. *Apiculture*—beekeeping—predates recorded history. From the tombs of ancient Egypt, where archaeologists found a jar of honey—still liquid and with its characteristic scent after 3,300 years—to the skeps depicted in medieval art and all the way to the present, when honey is a staple in every grocery store, the practice of beekeeping has endured. In fact, it is currently trending to the point where backyard beekeeping has been called "the new backyard chickens." Urban, suburban, rurban, and rural, hives are proliferating everywhere, and with them a whole new contingent of beekeepers. Some people just plain love bees.

Barbara Bartell, who has been keeping bees since the early 1980s, is one of these people. "When I was about eleven years old, there were some bees in an old cedar tree limb. I can still see that tree and know exactly where it is. I used to go watch those bees and sometimes spread a little honey onto the limb. I like just being around bees." Instead of running away wildly and flailing their arms to rid themselves of a solitary honey-maker, people like Barbara seem to have an affinity with the golden insects, walking among buzzing workers with affection. There are even those who work hives without protective clothing, developing an almost mystical symbiosis. And then there are a lot of people who fall in between the two extremes and, affinity or not, are considering beekeeping for fun and/or profit.

Honey Bee Basics

There are about 250 kinds of bees in the world, ranging from various types of bumblebees to tiny solitary bees, with many others in between. Honey bees comprise a very small segment of the whole—just seven known species. What most people in the Western world call honey bees are members of the species *Apis mellifera*. There are, however, six additional *Apis* species, all of them making up three subgenera (subgroups). In the subgenus *Micrapis*, we find *A. andreniformis* and *A. florea*. These dwarf Asian honey bees build their nests in shrubs and trees, and people gather a small amount of honey from them. The subgenus *Megapis* contains *A. dorsata*, the giant honey bee, a fiercely protective species that builds on tree limbs or on cliffs and whose nests can reach more than 3 feet wide. This bee is found throughout southern and southeastern Asia. The third subgenus, *Apis*, contains the red bee from Borneo, *A. koschevnikovi*; *A. nigrocincta*, found in the Philippines; *A. cerana*, the eastern honey bee (in several regions of Asia); and our familiar *A. mellifera*, the European (western) honey bee. Only these last two are known to be kept in man-made hives.

Beekeeping has become popular on farms as well as in suburban or urban neighborhoods.

Most other "honey" bees are hybrids or subspecies of *A. mellifera*. For example, the African honey bee is actually a hybrid between an African subspecies, *A. mellifera scutellata*, and *A. mellifera*. Mankind has been crossing various strains of bees throughout history, many of which have their own names yet cross freely with other *A. m.* bees to further hybridize, making subspecies rather difficult to distinguish from each other. For example, the Russian bee is actually a strain of Carniolan bee, and the Buckfast strain is a cross between Italian and European dark bees.

As is the norm with any breeding program, beekeepers choose certain strains for their prominent characteristics. The European black bee tends to sting without provocation, so, as you can imagine, it is cultivated less frequently than it used to be. Barbara shares, "Most of the bees we have had are

The World of Bees

As comprehensive as I've tried to make this chapter, I've barely scratched the surface of beekeeping. The keeping and care of bees is a whole world of its own, a calmer world, a world in tune with the earth, the sky, and the elements of nature. It does have its challenges, but, as most beekeepers will confirm, the rewards make up for the difficulties.

The honey bee,
Apis mellifera.

The African
honey bee.

of the milder temperament, but one time we did have black bees, and they were hot tempered. Oh, my!" The Caucasian bee tends to make more propolis (bee glue) than other strains, and this has made them less popular in recent times.

Historically, honey production and overwintering ability were the two major traits cultivated by apiaries. Nowadays, with mystery maladies such as colony collapse disorder (CCD) still unsolved (see sidebar on page 209), factors such as parasite resistance and disease resistance, as well as ability to withstand other environmental stresses, are sought after as well. In truth, every strain of bee is important as beekeepers and scientists alike search for genes that will save these species.

To Bee or Not to Bee

The idea of keeping bees may seem like the perfect enterprise—increased pollination; sustainability; your own honey to eat, give away, sell, or make products with; beeswax—what's not to love? There are a few aspects you should ponder, though, before launching such an endeavor.

Fear factor. Can you tolerate flying, crawling, potentially stinging insects? If at all possible, experience what it is like to be surrounded by bees, having them land and crawl on you ten, twenty, thirty, or more at a time. Not everyone is cut out for it.

Time. The bees themselves don't take a lot of your time—being busy themselves, they are quite self-sustaining. However, in working your bees (performing necessary tasks in the hive), timing is important. You will need to schedule those tasks on your calendar and perform them if you want your hive to stay put (not swarm) and survive things like winter, parasites, and disease. You must also take some time before starting your hive to research and familiarize yourself with the art of apiculture.

Expense. Thankfully, beekeeping is not outrageously expensive. In fact, you can start up without a major financial outlay. It's the cumulative effect that can take a toll on your wallet. Bees aren't cheap to purchase, and, unless you're unlike every other apiarist, you'll want more than one hive and will be replacing your colonies periodically. Also, unless you're a woodworker and like to DIY, you'll have to buy the supplies to create your hive(s), which will need regular maintenance.

Difficulty. Beekeeping has some aspects that are—well—not exactly easy. They sneak up on you later in the season, after your new hive has been happily humming along all summer, the honey supers are full, and you siphon off a bit as your reward for being so successful. In fact, your bees seem so healthy that you might end up skipping a couple of the necessary tasks that you should

As an aspiring beekeeper, are you ready for this?

perform as fall turns to winter. And then, come spring, when your bees should be emerging to tackle another season of nectar, they don't. Or you do everything by the book, and still something goes wrong. Mites? Fungi? Colony collapse disorder? Perhaps the greatest challenge for apiarists is keeping their bees alive from year to year.

Strength. You need a certain amount of physical strength to successfully maintain your bees. Barbara says, "You better have some muscles to have bees, because a 'super,'—a box that holds the larger frames—weighs at least sixty pounds when full of honey, and it must be pried off and lifted from the box under it from a bent position!"

Space and location. This is important. Having your hives nearby so you can enjoy them may at first be exciting and charming, but the bottom line is that it makes for a high ratio of bee-to-human contact. Bees may be synonymous with the positive aspects of honey and pollination, but they're also stinging insects, and a certain sector of the population is allergic to bee stings. In a rural setting, this may not be too much of a problem unless you or any of your family members fall into this category, but plunking down a hive or two in a more urban zone may not sit well with neighbors, especially if they have children or pets who spend time outdoors. Although it's true that honey bees are not normally aggressive, kids and pets don't seem to get the concept of "live and let live." And, even if they're not actually bothering your bees, they may unknowingly be blocking the bees' flight path between hive and nectar source. But it's not just your neighbors. It's you. You may love those little golden workers, but if you're fond of dining out on the patio, sunbathing, hanging out laundry, or working in the garden, the fact that you've unleashed thousands of bees to share your outdoor space may not be as romantic as you imagined.

When pondering the perfect site for your hives, avoid locations such as near the doghouse or chicken coop, next to the clotheslines, and along the neighbor's fence line. You get the idea. Minimizing contact is the goal. If you have livestock, watch out for these animals as well—goats will climb onto your hives, and cows and horses will rub on them and knock them over. If your pasture is the ideal site for your hives, fence them in. Bartell, whose current hive is in her horse pasture, uses a hay ring with a cattle panel laid on top to encircle the hive. In some areas, bears can be a problem as well.

Neighbors and local regulations. Speaking of neighbors, if you're not out in the country, zoning laws may apply. Don't just take someone else's word for it; do the research yourself so that you can ensure your compliance. Regrettably, sometimes, in spite of the fact that zoning laws may be friendly to beekeepers, your neighbors may not feel the same. Again, if you live near someone who is known to have allergies, do them and yourself a favor and postpone your beekeeping experience until you can be confident that you won't end up in legal trouble. Your bees may not act like you expect them to. If your hive swarms and disrupts traffic or chases your neighbor off his ladder, you could end up with a problem on your hands. As Winnie the Pooh so wisely observed, "You never can tell with bees."

The learning curve. It's steep. There's a lot to learn, and your bees aren't going to wait politely while you figure things out. Much of it is on-the-job training, which can, at first, be intimidating. You now need to know about your local weather patterns, climate, and freeze dates—things you may not have thought about too much before. Weather is a silent but critical factor for bees. If it's too cold, they may not make it through the winter. If it's too cold or wet in the summer, they cannot leave the hive to gather nectar to make honey, but they still must eat. If the weather doesn't cooperate, the bees won't have sufficient food to reproduce and survive the winter unless you feed them.

Then there's the world of parasites, pests, predators, disease, bacteria, and fungi. Local agriculture, bee-friendly plants, and wild pollinators become more than just points of trivia. With everything you need to learn, it can feel like you're taking a crash course in meteorology, biology, chemistry, physics, and animal science at first, but hang in there. Beekeepers are long-distance runners, not sprinters. Sometimes you lose a hive in spite of all your efforts. It happens to experienced as well as inexperienced apiarists, and if it occurs several years in a row without your being able to ascertain why, you can feel like giving up. Just remember that everyone starts at the beginning. It's those who persevere that deserve to wear the title of "beekeeper."

Like so many before her, Barbara Bartell is a great example of someone who started with no prior experience and just kept at it. "We lost our first hive during our first year. Bought two more the next spring." She has kept bees—sometimes one hive, sometimes several—for more than

You have to provide your bees with the right conditions for them to do their job.

The presence of bumblebees and other pollinators indicate healthy biodiversity.

thirty years, and, with each passing year, she has become savvier.

Apiculture is beekeeping, but the reality is that bees cannot be "kept" in the true sense of the word. They may live in a hive that you own, but they are wild creatures that know no boundaries and no owners. Honey bees will forage up to three miles from their homes. If most of that territory is prime for bees—meadows rich in flowers, uncultivated countryside, hedgerows, and organic agriculture—congratulations. You might be able to keep multiple hives if you so desire. Don't underestimate the importance of improving the habitat for bees. Selecting bee-friendly wildflowers and plants or letting some of your garden or property go a bit "wild" will encourage honey bees to thrive close to home.

Pollination

We learned all about pollination—the process of pollen being distributed among blossoms of like kinds, primarily by insects and specifically the honey bee—in third or fourth grade. According to the National Resource Defense Council, more than $15 billion worth of crops are pollinated by honey bees every year, with an additional $150 million worth of honey produced. There are other pollinators—bats, birds, butterflies, and more—but honey bees have an advantage in that they can be somewhat regulated by moving large numbers of them into close proximity to crops, orchards, and fields that need to be pollinated.

To produce honey, the bees must have blossoms, of course, so sufficient supply becomes of primary importance to the success of your hives, giving rise to the question of whether your neighborhood or farm can support the influx of ten, twenty, or thirty thousand more pollinators. Your bees will need enough nectar to produce sufficient honey stores to get them through the winter, and other bees and pollinators will be after the same flowers. Therefore, pay attention to biodiversity. Observe the activity of bumblebees and native wild pollinators. Their presence indicates a healthy ecosystem, and introducing a hive or three shouldn't cause an imbalance. But if bumbles are rare visitors to your farmstead, you may not have sufficient nectar sources to support an influx of honey bees. An exception might be in areas where pesticides have decimated wild pollinator populations; in that case, though, keep in mind that pesticides kill the good and bad insects with equal aplomb.

Unfortunately, many agricultural areas are still far from organic, and the use of various pesticides can decimate your hives. Even if your bees survive, there are growing concerns that insecticides and herbicides that are systemic in the plants will show up in the honey and also may be contributing

Honey bees will range up to three miles away from the hive to find pollen.

to CCD. Largely due to these environmental issues, some beekeepers in more urban areas are discovering that their bees are actually more productive and healthier than those near arable land.

Cost

Cost is, of course, also a factor when considering beekeeping. The good news is that once you make the initial investment, costs decrease (unless you have to replace the bees). Expenses fit into three main categories: start up-investment, bees, and equipment.

Start-Up Investment

Initially, you'll need a hive and all of the components that go into it. Rule number one: Never opt for a hand-me-down hive acquired from someone else. Although this may appear to be a cheap way to launch, unless you know why the owners are passing it along and how it was used, you may

be acquiring a whole raft of problems. Drawn wax may contain disease or other contaminants from a previous colony, as might other components. It's much better to start fresh and give your bees the optimal chance for survival.

There are various models out there, but by far the most common is the Langstroth hive. For DIYers, the Michigan Beekeepers Association has downloadable plans at

Your initial outlay includes the cost of the hive and beekeeping accessories.

Parts of a Langstroth-Type Hive

By far, the most commonly used hive type is the Langstroth design, which is a square or rectangular hive box with its various components as follows.

Hive stand/base: You can set your hive directly on the ground, but this is not recommended because moisture can wick up from the soil. Keeping the bottom board off the ground not only prevents it from eventual decay but also makes it easier to keep the bees' entrance free of grass and weeds.

Bottom board: Serves as the bottom of the hive and as a platform for the bees from which to take off and land. These boards are available from beekeeping-supply retailers and are reversible, giving ⅜- or ⅞- inch opening option. Because of this opening, your hive should be slightly tilted forward to keep rain from running in.

Hive body: This is basically a wooden box without a top or bottom. The standard size provides room for ten frames and can be found in four different depths:

The full-depth, or deep, hive body is 9 ⅝ inches high and is most often used for brood rearing because it provides adequate space for large, solid brood areas. You can also use a deep hive body for honey supers, but, when full of honey, the structure can weigh more than 60 pounds, which is a lot of weight to lift and move.

The medium-depth box, also called a Dadant or Illinois super, is 6 ⅝ inches high. This size, although most convenient for honey supers, can't be cut efficiently from standard-sized lumber and therefore costs a bit more per square inch.

The intermediate size (7 ⅝ inches) falls between the full-depth and medium-depth boxes, and beekeepers who make their own boxes tend to prefer this size.

The shallow-depth box (5 ¹¹⁄₁₆ inches high) is the lightest to manipulate (about 35 pounds when full of honey) but the most expensive per square inch of usable comb space.

There is an even smaller super, called a section box or comb super, which measures 4 ⅝ inches high and is used for the specialized art of producing section comb, which is best left to more experienced beekeepers.

Frame and combs: Frames holding beeswax comb are the basic components inside a hive. In a man-made hive, the wooden or plastic frames are typically fitted with foundation (a sheet of beeswax or a sheet of plastic coated with beeswax, often embossed with worker-bee-shaped cells and ready for bees to draw out). Plastic foundation and frames are becoming increasingly popular. Beekeepers swear by their own preferences, so do your research and talk to beekeepers in your area to help you decide which type is best for you. After a while, you'll develop your own preferences. Just remember that only well-supported foundation results in well-drawn combs.

Thin foundation is used to produce chunk, cut-comb, or section-comb honey, whereas a thicker, heavier foundation is better for use in the brood chamber and in frames for producing extracted honey because the thicker foundation can stand up to the centrifugal force of the extractor.

New foundation should be introduced to established colonies only during a major nectar flow or when the hive is undergoing rapid growth (a package, swarm, or colony split) because it takes a lot of nectar and labor for the bees to create new comb.

Queen excluder: Without a queen excluder, the queen can access all parts of the hive and may lay eggs in the center of every frame. This means that you'll have to harvest honey from frames with brood, which you don't want to do. A queen excluder is basically a grill or grid of plastic or metal with openings big enough for workers to pass through but too small for the queen's oversized abdomen.

Only about 50 percent of beekeepers use excluders; the others feel that it affects honey production, stating that workers don't like to pass through the narrow openings to store honey until there is no room down below. If you decide to go with a queen excluder, minimize this problem by leaving out the excluder until the bees have begun to store nectar in the super. This nectar will entice the bees upward. Those beekeepers who don't use an excluder position a full honey super directly above the brood chamber to serve as a natural barrier to the queen, who is looking for empty cells in which to deposit her eggs.

Inner cover: This cover rests on the topmost super beneath the outer cover and prevents your bees from gluing the outer cover to the super with wax and propolis. It also provides insulating air space just under the outer cover. During summer, the inner cover helps protect the interior of the hive from direct sunlight; in winter, it helps prevent moisture-laden air from direct contact with the hive's interior. The inner cover's center hole can be fitted with a bee escape when you want to harvest a honey super.

Outer cover: This telescoping cover fits over the inner cover and the top edge of the topmost super and is usually made of wood covered with a sheet of metal to help withstand weathering and leakage.

Materials: Although hives are traditionally made out of pine, redwood, or cypress, today you can get all of your hive components in plastic. Sound second-rate? Actually, plastic is durable, lightweight, strong, and easy to assemble and maintain. That said, plastic does not breathe or allow much ventilation, so beekeepers tend to choose plastic frames but avoid plastic hive covers, bottom boards, and hive bodies. When used in these larger pieces, plastic can warp, and some plastics allow too much light into the hive's interior, which hinders the bees in drawing out the comb.

A Langstroth hive is the most common model used by beekeepers.

www.michiganbees.org/beekeeping/in-the-beekeepers-workshop. Although ten-frame hive bodies are ubiquitous, some beekeepers prefer eight frames because this size tends to approximate the size of a cluster of bees. Eight-frame boxes are often homemade, but some suppliers sell them as "English garden" hive boxes. Other apiarists will use a ten-frame box but put eight frames within, giving more space between the frames. Most beekeepers use either two full-depth hive bodies or one deep box and one shallow box for the brood area. One contingent advocates the use of three shallow boxes because the frames will all be interchangeable and their size/weight is manageable. This is, however, the more expensive route and requires more time overall because you would have thirty frames instead of twenty. Alternately, some beekeepers use two medium supers.

There are other hive models that leave smaller carbon footprints and render you less dependent on outside sources, although they may not yield quite as much honey. Most of these more "natural" options are variations on the top-bar type of hive as opposed to the conventional box. The advantages of this type of setup are the cost and the sustainability; there is a whole movement afoot promoting natural beekeeping. Find free downloadable plans for a top-bar hive from the Barefoot Beekeeper: www.biobees.com/build-a-beehive-free-plans.php.

Bees

Next, you'll spend money on the bees themselves. Unless you're fortunate enough to get hold of a swarm and transfer it to a waiting hive, you'll need to purchase a queen and her retinue. A typical starter hive can come in what is referred to as a "package"—usually about 3 pounds of bees along with a queen in a separate screened box. You must then transfer the bees and queen to the hive. Another common way to purchase bees is what is called a "nuc," a small nuclear colony containing three to seven frames on which a freed queen is already laying eggs and workers are already making comb and tending the brood.

A colony already up and running helps rule out some of the uncertainty of starting up. However, this option is a bit more expensive, and, in addition, there is a certain joy and excitement in picking up your package of bees from the post office, taking them home, and watching them transform an empty box into a functional hive in a few short weeks.

Equipment

Equipment is everything other than the hive and the bees that you need to make beekeeping and honey harvesting doable, including beekeeping gear, honey extraction and harvesting equipment, and all of those add-ons that tend to find their way into a beekeeper's operation.

Hive Organization

Let's discuss a quick overview of bee society to hit the important points. There is one queen. There are workers bees (females) and drones (males). Without exciting outrage or revolt due to discrimination (indeed, the bees seem quite oblivious to the imbalance), the workers do all of the work—making wax, feeding bee larvae, gathering nectar and pollen. In the summer, a worker may last only four to six weeks, whereas winter bees—ones that hatch after nectar gathering season— can live for months. The male drones, hatched from unfertilized eggs, are not able to fly very far, and they can't gather nectar or pollen for themselves. In essence, they do nothing but eat and wait for a chance to mate with the queen.

Although this doesn't seem like a fair division of labor, drones are essential because the whole hive depends on the queen and her ability to lay fertilized eggs. The largest bees in a colony (except for the queen), drones have no pollen baskets, wax glands, or stingers, and the queen herself determines when to lay unfertilized eggs, which will become drones. Drones stay in the hive for about eight days before taking their orientation flight. Since they eat up to three times as much as worker bees do, the population of drones is kept small—two or three hundred per season compared to tens of thousands of workers. Drones die after mating, and when cold weather reduces nectar and pollen stores, they are ejected from the hive to starve. The queen will lay new drones in the spring.

The worker bees perform all of the tasks essential to the colony.

Capsules containing royal jelly, to be fed to potential queens.

The queen starts out like any other worker bee. Her royal status and her ability to lay eggs is developed by the fact that she is fed copious amounts of "royal jelly," a special bee food. According to the Mid-Atlantic Apiculture Research and Extension Consortium (MAAREC), a queen can produce approximately 250,000 eggs a year (that's up to a million in her lifetime!) and can live from two to five years, laying an average of 1,500 eggs a day during peak production and tapering off to lower numbers during other times of the year.

You can tell the queen from the other bees because her abdomen is much larger and longer than that of drones or workers, and her wings cover about two thirds of that abdomen. She has neither pollen baskets nor functional wax glands, and, although she does have a stinger, it is shorter, more curved, and less barbed than a worker's. Her genetic makeup, along with that of the drone with which she mated, contributes significantly to the temperament, size, and overall quality of the colony.

A newly hatched bee emerging from its cell.

Honeycomb with capped brood.

In addition laying eggs, a queen is important to the hive because she produces pheromones—chemicals that serve as a sort of social "glue," unifying the hive and giving it certain characteristics. These pheromones extend up to 3 miles from the hive, hence the radius of area available to worker bees out on gathering forays. A decrease of these pheromones signals that a queen is failing. When worker bees detect the lessening of the queen's chemical influence, they begin to make new queens to supersede her (a beehive is a very pragmatic world, you know).

Worker bees, which are sexually undeveloped females, hatch from fertilized eggs and have scent, brood food, and wax glands along with pollen baskets and stingers, all of which equip them to do the work of the hive. When workers first hatch, they build comb from wax that they produce, clean and polish the cells, take care of the queen, feed the brood (growing larvae), process incoming nectar, guard the hive entrance from intruders, remove dead bees and other debris from inside the hive, and cool the hive through "air conditioning" (fanning their wings). Later, these workers "graduate" to the work of field bees, foraging for pollen, nectar, propolis (plant sap), and water. No wonder their average lifespan in the summer is only about six weeks!

Interestingly, if a colony loses a queen unexpectedly, the ovaries of several worker bees automatically develop, enabling them to lay unfertilized eggs. This phenomenon is believed to be prevented during normal circumstances by the presence of brood and the queen's pheromone chemicals. Laying workers usually indicate that the colony has been queenless for one or more weeks, but they do not actually aid the hive in replacing her because all eggs laid by workers are drones. Sometimes, laying workers will be found during swarming conditions or when a poor queen is at the head of the colony. A colony with laying workers is easy to identify because they lay more randomly within the brood combs, sometimes from five to fifteen eggs per cell, and they do not place the eggs in the bottom of the cell as a queen would do. Eggs from laying workers don't always hatch; if they do, and if laid in the smaller cells, the offspring do not survive to maturity.

Queen, worker, or drone, all three types go through the same developmental stages of egg, larva, and pupa, which is referred to in the beekeeping world as "brood." Developmental time varies, however. Drone larvae are fed honey and pollen. Worker larvae receive a certain amount of royal jelly along with their mixture of honey and pollen, and the queen, as previously mentioned, consumes large amounts of royal jelly. Once she has mated, she begins to lay eggs, one per cell. Each egg is attached to the bottom of its wax cell and looks like a tiny rice grain. After three days, the egg hatches into a grub, starting the larval stage of development. Healthy larvae curl in a "C" shape and are glistening pearly white. Worker cells are capped at five and a half days, queen cells at six days, and drone cells at approximately six and a half days.

Brood comb becomes darker in color than the golden honeycomb.

Before capping, the larvae are fed by nurse bees and grow lengthwise in their cells. Remaining the same healthy white, they spin thin cocoons. Once the cells are capped, they enter the prepupal stage, in which they begin to assume their adult forms and gradually take on color. Workers emerge twelve days after capping, queens emerge seven and a half days after capping, and drones emerge fourteen and a half days after capping.

A good queen lays healthy brood, which you assess by looking at the capped brood cells, which should have a solid pattern with only a few cells missed. Furthermore, capping should be convex, unpunctured, and medium brown in color. Drone brood is usually found along the edges of the brood comb. MAAREC notes that due to the variances in developmental time, to maintain a sufficient amount of workers, "A good ratio should be four times as many pupae as eggs, and twice as many as larvae."

Over time, brood comb becomes dark and is easily distinguished from honeycomb, which retains its characteristic golden hue. One method of hive management advocates using a single full-depth hive body, which theoretically gives the queen all the room she needs for egg laying. In reality, a beekeeper uses a single deep super when he or she wants to crowd the bees and cause comb honey production, when a package of bees is installed, or when a colony is first established. Otherwise, additional space will be needed for food storage and maximum expansion of the brood nest.

Dead Bees

There will be dead bees—sometimes what seems like a lot of them—and there's little to nothing you can do about that. Periodically clear them out of the opening and be comforted by the fact that many of these are workers from the summer months that have reached their maximum life spans. Do not open the hive until spring has sprung.

As the autumn leaves appear, the bees begin to prepare for survival during the colder weather.

Bee Seasons

The honey bee's year starts not in January but in September, and the bees' activities from September to January greatly influence the survival of the colony. When nectar and pollen sources dwindle in the fall, this is reflected in diminished brood laying and hive population. Old bees die off, but young ones survive the winter. During this season, workers use propolis, gathered from tree buds, to seal all cracks in the hive and reduce entrance size to help keep out cold air. Workers begin to oust drones, denying them re-entrance to the hive to reduce the overall demand on food stores. At 57 degrees Fahrenheit, bees form a tight cluster in the center of the hive, where the brood is, to keep the next generation at approximately 93 degrees Fahrenheit. The queen may stop laying during October or November in colder regions; in milder areas, egg laying and brood rearing usually never completely stop.

The colder it gets, the tighter the bees cluster, expanding and contracting as temperatures rise and fall. During warmer spells, bees within the cluster shift to allow those on the outside to cycle in where they, too, can access honey stores. Sadly, in a prolonged cold spell, those workers on the outer layer may end up starving even though they are only inches away from food. In colonies with good honey and pollen stores, workers will begin to stimulate the queen through special feeding so that she begins laying during late December or early January, even in colder climes.

This new brood will help replace the bees that die during the winter, and the extent of this early brood is directly connected to the amount of pollen stored the previous fall. If the colony is low on pollen, brood rearing is delayed until spring, when workers can bring fresh pollen. These hives tend to emerge from winter with smaller populations, and colonies with ample pollen and honey supplies and plenty of young fall-hatching bees usually result in strong hives come spring.

Spring

As days lengthen and workers start bringing in new pollen and nectar, brood rearing goes into high gear. The drone population will be minimal to nonexistent, whereas the worker bee population explodes to boost food gathering. In fact, the workers may bring in so much that surpluses of honey and/or pollen begin to accumulate (good news for beekeepers!). Workers also bring water into the hive to dilute thickened or granulated honey.

As the colony expands, the nest area becomes crowded, and more bees become visible at the entrance of the nest. If you observe bees hanging in a cluster around the entrance on a warm day, you'll know that your hive is overcrowded. Under these conditions, the queen increases drone laying to prepare for the colony to divide by swarming, and, while she's busy laying workers and drones, workers also prepare to rear a few new queens. MAAREC states that during times of emergency, swarming, or *supersedure* (replacement of a failing queen), workers enlarge several cells in which the queen has laid female larvae and begin to feed those selected larvae royal jelly at a heightened rate (individual colonies and different races/strains of bees produce varied numbers of queen cells). Interestingly, queens raised to supersede an old queen are often better than those produced during emergency situations because they have the advantage of time and are fed more royal jelly than those raised during the loss of a queen or when the colony becomes overcrowded.

Colony Collapse Disorder

The US Department of Agriculture (USDA) Agricultural Research Service reports that CCD (also called crown colony collapse) first showed up in 2006, when beekeepers started reporting losses of 30–60 percent of their colonies. This is not the first time that apiarists have been faced with unexplained losses; scientific literature mentions honey bee disappearances in the 1880s, 1920s, and 1960s as well as in 1902 in Utah's Cache Valley and in 1995–1996 in Pennsylvania. There's no way of knowing if these events, too, were caused by CCD, but the descriptions sound similar.

The main symptom of CCD is a very low or nonexistent adult honey bee population in the hive. However, in a sort of mysterious twist, you will see no dead bodies, and the hive will still have a live queen, honey, and brood.

In 2007, Barbara Bartell lost her hive to CCD. "It's a strange situation," she recounts. "I had gone out about a week before, and the bees were OK. A week later, I went back, and there were no bees alive and very few bodies lying around. There was plenty of honey—at least half of the frames had honey—so it was not a lack of food."

Often, but not always, varroa mites—virus-transmitting parasites of honey bees—are present in hives hit by CCD. Whether they cause the disorder has not been proved. Ultimately, researchers think that CCD may be caused by a combination of factors falling into the categories of parasites, pesticides, herbicides, pathogens, environmental stressors, and management stressors. Some cities, including Seattle, have banned neonicotinoids, pesticides related to nicotine, and huge die-offs in Canada and other regions are bringing the CCD problem front and center. While the jury is still out as to a concrete cause, beekeepers can hedge their bets by caring for their bees' overall health and habitat, countering parasites and pathogens, feeding during nectar scarcity, and addressing known mortality factors.

When you see your colony building these queen cells, you'll know that they're preparing to swarm. While this is going on, the crowded hive will seek to expand by building out new comb if they have room. They will generally store honey in the new comb, using older comb for brood rearing and pollen storage. This is the time to intervene to prevent the eminent swarm. MAAREC links the prevention of swarming to sufficient room in the hive for brood and adult population, sufficient queen substance (pheromones), and the hive environment as it relates to these factors. Beekeepers achieve this through providing several components:

- Plenty (not just adequate) of room for egg laying. For strong colonies, two hive bodies devoted to brood may not be enough. Rotating brood frames every six to eight days can help; also ensure that brood comb is in good condition, with a minimal amount of drone and honey/pollen cells.
- Plenty of space for nectar storage.
- Exposure to maximum sunlight early in the season and to shade as summer temperatures rise. A south-facing entrance will catch the morning sun, and afternoon shade is recommended. White hive bodies help control summer temperatures.
- Sufficient hive ventilation. Be sure to remove winter restrictions to the hive entrance, but do not allow too large of an entrance because invaders may view that as an invitation.
- A young queen of a low-swarming variety.
- Removal of queen cells. If the cells are advanced or completed, their removal will only postpone, not prevent, a swarm. In this case, you'll need more drastic measure to head off swarming. MAAREC has a very helpful page about this process on its website (https://agdev.anr.udel.edu/maarec). If the hive is just beginning to form queen cells, providing more room and removing the cells will help prevent a swarm. After about ten days, check again to make sure that new queen cells have not been started.

A honey bee swarm hangs from a tree.

Swarming

If you've ever seen a swarm of bees on the move, you know that it's a wondrous thing. In Barbara Bartell's words: "I am always fascinated when I hear a bee swarm, like I did earlier this year. I heard this huge sound—it sounded like a motor but I knew that it wasn't—and then it dawned on me: it's bees swarming! I looked, and, sure enough, they were swarming by the chicken house. The air was full of them, but they settled onto a branch of a tree that reaches over the chicken house, and there was no way could I get them."

That swarm took off about an hour and a half later, but in July of the same year, Barbara heard yet another swarm. This one settled on the grapevines in her garden, and, with the help of a friend, she was able to transfer the cluster to a hive. This is now her current colony and is doing well.

If you take up beekeeping, it's a great idea to have everything you need to set up a hive on a moment's notice because swarms don't stick around for long. In fact, as Barbara witnessed, if you see them, they can see you. They may up and fly away while you're getting your equipment, so don't get too close to them until you're ready to capture the swarm. If you can secure a swarming colony, you'll have your hive up and running in no time, and the bees will be cost-free! Swarming usually takes place before the main nectar flow comes on. In southern, western, and central states, this is usually from March to June, but swarming can happen anytime a hive gets overcrowded between April and October.

According to MAAREC, as new queens develop within their cells, several fascinating changes occur in the hive: "The [current] queen loses between one third to half of her normal body weight. The field bees do less work and may congregate at the hive entrance and/or on lower frames. More drones are often reared; fewer and fewer eggs are laid by the queen. Finally, prior to leaving the hive, workers gorge themselves on honey and nearly cease normal flight activity." Then, on a warm day, the old queen and roughly half of her bees will swarm out of the hive, fly around for several minutes, and then come to rest on a tree limb or some other object. Workers cluster around their queen to keep her safe, waiting for scouting bees to find a new home. When those scouts return, they dance on the cluster, and when all of their sisters get the message, the cluster breaks up and flies to their new location. Upon reaching it, the workers begin to make comb and gather food, and the queen begins to lay brood. Just like that, the new colony (approximately half of the old colony) is up and running.

Meanwhile, back at the old hive, field workers keep gathering pollen, propolis, water, and nectar. Hive workers care for the brood, guard the entrance, and make comb. They nurture emerging drones so that there will be consorts for the queen when she is ready to fly. When the first new queen hatches, she first eats some honey, grooms herself, and then hunts out all of the other queen cells. It may seem brutal to us, but somehow she knows that a colony can have only one queen, and those that have not yet hatched, she kills in their cells. She engages any that have already hatched in mortal combat. One queen emerges victorious, and when she is

A beekeeper in protective gear, preparing to capture a swarm.

about a week old, she flies out to mate with one or more of the drones. These males then die, but she returns. At this point, nurse bees begin caring for her, and she will begin to lay eggs within three or four days. This is how it works if you, as a beekeeper, let nature take its course.

Capturing a Swarm

Letting your hive swarm is a bit risky—you may gain a whole new colony, but you can just as easily lose half of your bees because they don't send out advance invitations to come and watch the show. It's much preferred to expand your hive(s) on your own timing, but, in the event of a swarm (yours or some other colony's), capturing it is not difficult as long as the bees cluster somewhere that you can reach. Here are a few steps toward success:

- As previously mentioned, if you see a swarm, don't approach it until you're ready to capture it because the bees may just leave while you're suiting up in your protective gear and gathering your equipment.
- Spray the cluster down with scented sugar water (vanilla is nice). Wet their wings and get them thinking more about eating than about swarming. Alternately, you can use a smoker, which works to keep bees calm.
- Clip away any small branches that could hinder the capture.
- Find the main limb and hold your capture box under the swarm. Ideally, your capture box should be a nuclear box, which is a mini-hive to which you can add frames. Remember that these bees have gorged on honey and will be heavier than you might expect. Also, as Bartell puts it, "They are not dangerous when they swarm. It's like their attention is on keeping together, protecting their queen, and looking for a new home."
- Reach up and clip the main limb so that it falls into your capture box and then set the box down under the spot where the swarm had landed (the scouts will be returning soon to look for their colony).
- Reach in for the branch, gently shake all of the bees off into the box, and then lay the branch in front of the entrance. It will smell like the queen, and the scouts will be looking for it. Those pheromones again!
- Gently set your frames down into the box, letting them rest on the cluster of bees at the bottom (the frames will sink down as the bees move up onto them).
- Step back and watch the bees take ownership. They'll be as thrilled as kids at a birthday party, climbing up and down the frames. The more the queen spreads her pheromones, the more at home the colony will feel.

Painting the Hive Parts

You should paint all of the wooden hive parts that are exposed to weather to preserve the wood. Use a good latex or oil-based *exterior* white (or pastel) paint. It may be tempting to use up any old paint that you have around the house, but remember that a light color helps prevent heat absorption during the summer. If you have multiple hives near each other, painting different shapes or letters on the front of each hive will help bees recognize their colony and reduce the tendency of bees to "drift" between colonies.

At this point, leave the hive cover off, setting it down on the box at an angle so that it contacts on four points, and give the bees a jar of sugar water on a porch feeder so they can begin to feed right away. (Bartell advocates setting the syrup jar on a rag so the bees can sip without drowning.) If the queen is inside the hive, you'll see them sticking out of the entrance, fanning hard, their back ends pointed skyward, as they release scent to call their scouts to their new home. Once you see this happen, you can put the lid on and leave them alone.

Well after sundown, return to the hive box, plug the opening, and move it to its permanent location, which should be 80–150 feet away from the hive that they exited, or they may just swarm again. Leave the plug in overnight and remove it early the next morning.

Don't bother the bees for a couple of weeks as they settle in. If you have other hives, one way to ensure that the new comb will stay put is to take a full brood comb (brush it free of bees) and insert it into your new hive. With brood to tend, the bees will lock into their new location. You may continue to feed them sugar water for a couple of weeks to give them a solid start.

Summer

No matter how hot it gets outside, bee colonies must hold the internal temperature of their hives to approximately 93 degrees Fahrenheit. Workers do this by gathering water and then spreading it on the interior of the nest to evaporate. During summer, the colony reaches peak population and peak food-gathering and storage activity for the upcoming winter. Maximum daylight affords maximum hours for gathering. Unless excess rain, cold, or drought keeps workers home, this is the time when beekeepers may remove a portion of honey, leaving sufficient stores for the hive to overwinter.

So, what is sufficient? Good question, and not an easy one to answer. Climate and weather, hive structure and amount of ventilation, number and kind of bees, and the ratio of warmer days to

Painting beehives white helps keep the hives from getting too hot.

abnormally cool ones are some main issues. Each colony's need will vary somewhat, so the best thing to do is estimate on the high side. Don't make the mistake of assuming that if your honey supers are full, the lower brood boxes will also carry sufficient honey stores. Check to make sure, or else you may be leaving your bees practically without food to overwinter. Some general guidelines are as follows:

- Bees in the southern climes may thrive on as little as 40 pounds of honey, bees in the middle states require about 60 pounds, and northern bees may require 80 or 90 pounds on average, though even the word "average" is a bit vague. Therefore, in all but the warmest areas, beekeepers can assure a good supply of winter honey by leaving 80–90 pounds per colony, which should eliminate the need for supplemental feeding of sugar syrup.
- A full ten-frame deep box weighs 80–90 pounds. A full ten-frame medium box weighs 65–75 pounds. Discounting the weight of the structure and dividing by ten, a full deep frame holds about 8 pounds of honey, and a full medium frame holds about 6 pounds. However, if you use eight or nine frames in a ten-frame box, each frame will weigh slightly more.
- Ideal fall brood will be in the center of the lowest box, flanked by frames of honey and pollen, with the outermost frames filled with honey. The second deep/medium box will be filled with honey. For northern areas, twelve deep frames completely full of honey should give you about 96 pounds plus any stored on the pollen frames. In more temperate areas, using a medium box yields about 72 pound of honey.

Honey

A worker bee sucks flower nectar into her honey sac (a secondary stomach). Amazingly, one worker can hold .70 milligrams of nectar—the product of 1,500 blossoms—in one load! She will carry it to the hive, where younger workers suck it out of her honey sac and mix it in their own stomachs with an enzyme called *invertase,* which breaks the nectar's sucrose into glucose and fructose to make it digestible. After converting the sugars, they regurgitate the nectar into honeycomb cells and then fan with their wings until the moisture content reaches less than 18 percent. This low-moisture honey is virtually impervious to spoilage and, if stored correctly, will last practically forever (remember the Egyptian tomb honey?). Be sure, though, that when taking honey from the hive, you do not take it from uncapped cells because the higher moisture content makes it susceptible to spoilage.

Working the Hive

Take your time when you're working bees and keep in mind that they don't actually want to sting you—if they sting, they die. Although some strains of bees or individual colonies may be "hotter" than others, bees generally just want to go about their business. If you can avoid abrupt movements (like swatting at them) and avoid crushing any bees, chances are they'll adapt to your presence without making too much fuss. As a general rule, work your hives around noontime on a warm day. Avoid late afternoons or stormy or windy days whenever possible because the bees may be more agitated at those times or under those conditions. Also make sure that your neighbors aren't outside sunbathing or playing. Here is a basic rundown of how to "work" a Langstroth hive.

- Light a fire in the bottom of your smoker with a small amount of newspaper, cardboard, or something else that ignites easily. Slowly add your fuel (tree limbs cut into small pieces, wood chips, leaves, pine needles—choose a fuel that burns cool). Pump the bellows often and have more fuel on hand to use when you need it.
- Don your suit and hat with veil. Although you may wear gloves as you get used to working with your bees, it's best to wean yourself off of them as soon as you feel comfortable. Without gloves, you can feel the bees much better and avoid crushing them; crushing them causes them to sting and causes the rest of the hive to feel threatened, become irritated, and start stinging as well.
- Approach your hive from the rear or side, reach around to the front, and puff a bit of smoke at the entrance. Next, lift the lid and puff a bit of smoke underneath. If the lid is really glued down with propolis, tap the corners lightly with a rock or hammer to dislodge the seal. Keep in mind that bees don't appreciate being disturbed, so keep the hammering to a minimum.
- Remove the top cover and lay it on the ground upside down.

A bee smoker is a helpful tool for working your hive.

A box with its cover off and frames taken out.

An apiarist uses a brush to gently remove bees from a frame.

- Using a hive tool, pry off the inner cover, puffing a little smoke under it as you do so. Lean it up against the side of the hive where you won't accidentally bump it or knock it over and crush some bees.
- Now you're ready to examine the frames one by one. Give a puff of smoke to encourage bees to move off the frames. Cut through any "burr" comb—comb that is between the frames—and, working carefully so as not to crush any bees, pry loose the ends of the frame second closest to you (the outside frame may be connected to the hive wall with burr comb, so it's not the easiest to remove first).
- Once the frame is loose, carefully lift it straight up out of the super, watching for the queen, who could be crawling on it. This is more likely if that frame happens to have a lot of open brood cells. You'll be looking for things like queen cells, how healthy the brood pattern is, how many drones cells are present, and diseased brood, as well as checking the levels of honey and pollen.
- Check the other side of the frame by tilting it so that the top bar is vertical and then flipping it over so you can see the other side. When you are done examining the frame, lean it against the hive stand.
- If it's a cool day and you've pulled brood comb from the hive with the frame you removed, try to get the brood comb back into the hive in less than two minutes so the brood doesn't cool down and die.
- Continue with each frame as previously described, with the difference that you won't need to set subsequent frames outside the hive. Merely set each one back down into the hive after examination. Give the bees a chance to move out of the way before you lower the frame all the way back into position.
- Finish by ensuring that all frames are back in their original positions and placing the removed

frame back in the hive. Eight frames in a ten-frame box would be the minimum. Bartell advises, "If you don't keep the box full of frames, the bees build comb between the frames, and it's a mess to deal with!"

Keep your smoker handy throughout all phases of working the hive.

- Next, pry the box loose from the lower super, lift it carefully (it can be quite heavy!), and set the whole box onto the lid that you laid upside down. Set it at an angle so the box contacts the lid on only four points in order to minimize the possibility of crushing bees.

- Using smoke as needed, work the bottom hive box in the same manner as the first. It won't be necessary to remove this second box unless you have yet another one below it.

- To reassemble, puff some smoke under the super to encourage the bees to move up and off the bottom edge. If the super is very heavy, set its front edge down on the back edge of the lower box and then slowly slide it forward. The bees will move out of your way. When you get to the last quarter inch or so, give the back edge and then the front edge a quick swipe with your brush to remove any bees that might get crushed and finish sliding the super into place. For a lighter super, hold it half an inch above the lower box and, with a slight circular motion as if you were wiping a spot off the kitchen countertop, lower it down to rest on the box beneath. Somehow, that motion causes the bees to move out of the way.

- Replace the inner and top covers and walk away!

Honey Time

When removing honey supers from your hive, rid them of most of the bees with one or both of the following methods. Method 1: Insert a "bee escape" into the hole in the center of the inner cover and then lift the honey super and put that inner cover with the escape between the honey super and the lower levels of the hive. The bees will escape from the upper super to join their sisters. This takes a bit of time.

Inspect frames one by one before replacing them.

Collect the honey as it drains from the comb.

Method 2: Drive the bees downward by using a liquid chemical (such as Bee Go or Honey Robber) sprayed on a fume board (a common accessory available from beekeeping suppliers); you then place the fume board like a cover on top of the honey super. The pungent odor forces the bees downward in just several minutes; in fact, if you leave it on too long (more than about five minutes), it can drive the bees from your hive completely, which you obviously do not want. Most of the bees in the super should be out when you take the super from the hive, but if they aren't, carry the super away from the hive and then shake or brush the remaining bees off of each frame as you pull it from the super.

Take your frames/super to a secure place that honey bees can't access, such as your kitchen, basement, or garage, because they'll sniff out the honey and move it back to their hive in short order. Comb often breaks when removed from the hive, and honey begins to drip out. Setting the super into a plastic garbage bag will help keep the leaking honey inside the bag rather than all over your kitchen and will also prevent bees from getting to it.

During your first year or so of beekeeping, you shouldn't expect a great amount of honey from your beehives. You might end up with a bit for your pantry and a few jars to give as gifts. At this point, you probably don't need to invest in honey extraction equipment. Following are three common methods to extract honey.

Strained Honey

If you have plastic frames and foundation, all you'll need to do is scrape the comb and honey from the foundation. Get a fair-sized container to hold the wax and honey and place a cloth filter (such as cheesecloth or paint strainer sacks from a paint store) over it.

Scrape the wax and honey off the plastic mid-rib and let it fall into the straining cloth that is draped over the container. The honey will pass through the strainer, and you can squeeze out most of what remains in the wax by wringing the cloth.

Rinse the comb with water and use this water to feed the bees. You can use the rinsed wax to make candles and a number of other products. If you don't have plans for the wax, just put it in a pan near your hive, and the bees will take it back, use any remnants of honey, and build more comb with the wax.

Set the frames back in the super and put it right back onto the hive where it came off, giving your bees a chance to clean up these frames, after which you can remove the frames and save them for the following season. Finish up by bottling your honey.

Cut-Comb (Chunk) Honey

If you use this method, opt for wooden frames and choose a wax foundation. The bees will draw out their comb on this foundation and fill it with honey. You then cut the foundation and honey out, leaving an empty frame. The easiest way to do this is to obtain a pan a little larger than the frame you plan on using. Many beekeeping supply companies carry suitable pans. You also need something to put the comb into, such as wide-mouthed jars with lids. Cut the comb to any size you choose and pour the extra honey drippings into the jar/container with the comb.

Extraction

If you get into bees long term, you may want to invest in a honey extractor and an uncapping tool (there are many models of both). The process is fairly simple: uncap the honeycomb, drop the frame into the extractor, and turn the handle. Centrifugal force draws the honey out of the uncapped cells onto the walls of the extractor, where it then drains down and can be tapped via a spigot at the bottom of the extractor. You can then put the frames back into the hive. Because the drawn-out comb is largely intact, you'll save your bees a lot of extra work, which will result in more honey stores.

A honey extractor in motion, with the frames spinning inside.

Preserving the Harvest:
Fruits and Vegetables

In all aspects of life, there's something quietly and supremely satisfying about reaping the fruits of what you've been laboring for, and this definitely applies to garden and orchard produce on the farm. All of that planting and weeding and tending will pay off. Harvest time! Exciting, yes, but until the produce is processed or stored, you won't have any extra time on the porch swing! When tomato plants are weighted down with luscious red globes, strawberries ripen in profusion, and sweet corn is ready by the bushel, it's time to shift into high gear. You'll be joining the ranks of generations before you as you freeze and preserve your bounty.

Storage Options

Many vegetables and some fruits store well for later use if you can control the temperature and humidity, thus root cellars are not necessarily a thing of the past. Carrots, beets, potatoes, onions, turnips—most root crops—can keep for quite a while in cold storage, as can certain kinds of apples or pears, but several factors are crucial for root cellar success. Consistent temperatures in the appropriate range, proper humidity, and ventilation to provide clean air and remove excess gases are all essential for root cellar storage. If you have a basement in your house, you may have the ideal place to construct a root cellar closet. The one absolute must for a basement to function in this capacity is that it remain dry—if yours is subject to dampness, it will not serve the purpose.

Carrots, potatoes, beets, and red onions are among the root vegetables that fare well in cold storage.

An ideal situation for a basement root cellar is a northwestern or northeastern corner with a window (away from direct sunlight). Partition off the corner with insulated walls and an insulated door. Remove the window glass and place metal screening over it to keep critters out and then install a flue board with an air intake flue to bring cool air in and an air outlet flue that allows hot air to escape. You can make flues out of wood or PVC pipe. Open and close both flues with a damper (parts are usually available at local hardware or stove stores). You will quickly get the hang of controlling the temperature by manipulating the flues. If you need additional humidity in a root cellar, simply add a pan of water.

If you don't have a suitable basement, other cold storage options include buried barrels or metal garbage cans, straw bales, or drain tiles. Temporary pits or buried piles also work, although they are less convenient and difficult to rebuild once opened, so you may need to use stored produce quickly after opening. A plethora of sources of information on cold storage are available through the Internet, libraries, and extension offices around the country.

The length of time that produce keeps in a root cellar depends on several factors: maturation period, storage conditions, and condition of produce. Late-maturing varieties harvested in the fall store better than varieties harvested in summer. Less-than-ideal storage conditions will shorten storage life. The crops that you store in a root cellar must be of the highest quality. Bruised or blemished fruits or vegetables are sure to go bad, and they tend to take others with them.

Some fruits and vegetables don't mix well in storage, even though their temperature and moisture requirements are similar. For instance, apples, tomatoes, peaches, pears, and plums all give off ethylene gas, which can cause potatoes to sprout. In addition, other fruits and vegetables can absorb the odor of strong-smelling vegetables, such as turnips and cabbage, so you should store these fragrant vegetables separately from other foods.

If you plan on storing your produce in a root cellar situation, if at all possible, harvest in the morning after the dew has dried but before the midday sun begins heating the crop because produce spoils more quickly if stored hot. According to *Mother Earth News*, all you need to do is brush off excess dirt, trim tops to 1 inch, cool overnight (refrigerating is great), and then store. Avoid washing root crops because it can remove a natural barrier that helps prevent spoilage. Likewise, produce spoils more quickly if stored when moist, so if you must harvest while dew is on the crop, make sure that the fruits and vegetables are dry prior to storing.

Preservation Techniques

Food preservation has been around for a long, long time. Preserving the harvest through techniques such as drying, canning, and freezing allows us to enjoy our bounty of summer throughout the year, and there is something special about rows of beans, tomatoes, or pickles in their pristine canning jars. They almost shout, "Let the winds of winter blow! We are prepared!" For a wide variety of recipes and food preservation information, visit the National Center for Home Food Preservation at nchfp.uga.edu/index.html. Funded by the Cooperative State Research, Education, and Extension Service (a branch of the US Department of Agriculture [USDA]), the center addresses food safety concerns for those who practice and teach home food preservation and processing methods and is operated by the University of Georgia Cooperative Extension Service.

Turnips are a strong-smelling vegetable that should be stored separately.

Another website to visit is www.freshpreserving.com. This site is a service of Jarden Home Brands, the manufacturer of home canning supplies, including Ball and Kerr jars. It has a good basic overview of canning techniques and a recipe search engine, along with a wide range of canning equipment and supplies for purchase, including the *Ball Blue Book of Preserving*.

Drying, freezing, canning—each method has advantages and disadvantages, and the crop itself indicates which methods give optimal results. Many people, used to the frozen food section of the grocery store, choose to freeze the bulk of their garden harvest. Some vegetables and fruits do well with this method, and others not so well. Whichever method you choose, pick blemish-free, fresh, firm, and ripe fruits and vegetables for preservation. Overly ripe produce should not be used (except when making fruit leathers).

Blanching Fruits and Vegetables

Most vegetables and some fruits require blanching prior to freezing and sometimes before drying or canning. Skipping this step can result in a disappointing end product, which can be so disheartening after all the time and effort you've put into planting, growing, harvesting, and preserving. *Blanching* means to briefly heat vegetables/fruits in boiling water or steam to inactivate naturally occurring enzymes that can cause undesirable changes in the foods, including nutrient loss, toughening, and loss of flavor and color. The brief heating also reduces the number of microorganisms on food, thus reducing the chance of spoilage.

Blanching is simple. You can either dip vegetables/fruits in boiling water or steam them; the latter takes a little longer, but it preserves more nutrients. The aforementioned

Stainless Steel

Use stainless steel or enamel pots for boiling and preparing foods for canning; avoid using aluminum or iron pots to blanch produce because the metals react chemically with the foods, leaching the minerals from the metal into the food.

Blanching and then freezing is the recommended way to preserve brussels sprouts.

websites or a book dedicated to canning will contain much more detailed instructions on preparing your produce for blanching, but, simply put, you need to clean, and in most cases cut, vegetables/fruits before blanching. To blanch in boiling water, bring the water to a boil in a large kettle. You will need about one gallon of water for each pound of vegetables or fruits. Submerge the vegetables/fruits in the boiling water using a colander, wire basket, or loose cheesecloth bag, or drop them directly into the water. For the latter method, remove them from the water after their allotted time using a slotted spoon for smaller produce and tongs for larger items.

To steam blanch, place one inch of water in an open kettle and bring it to a rolling boil. Use a steam basket designed for this purpose or a cheesecloth bag to suspend a thin layer of vegetables or fruits over the boiling water instead of submerging them, and then cover the kettle, allowing the steam to heat the produce.

The blanching time varies with the type of vegetable or fruit you will be processing. As soon as the time is up, remove the vegetables/fruits and immediately submerge them in ice-cold water. If freezing, chill the blanched produce until completely cooled and allow it to dry completely before freezing.

For items such as tomatoes and peaches, blanching is an easy way to deal with the skins. After three to five minutes in boiling water, the skins slip right off, making them ready to pack into jars and process.

Drying Foods

Drying is probably one of the oldest food storage methods. In fact, the Bible mentions drying food in the sun and smoking food over a hot fire, and archaeologists have discovered samples of foods they believe people dried in Jericho more than 4,000 years ago.

To successfully dry foods in the sun, you need very low humidity and long hours of very intense sunshine. A nice strong breeze, temperatures of 95 degrees Fahrenheit or higher, and humidity below 60 percent are perfect sun-drying conditions. In the eastern half of the country, this weather is rare, but those who reside in the southwestern United States have these conditions frequently. A solar dehydrator, a gas or electric oven, or a portable electric dehydrator offer options for drying food where nature doesn't provide ideal conditions.

Drying works well to preserve a variety of fruits, legumes, nuts, and beans.

Let's look at sun drying. To begin, place prepared foods on drying trays, making sure that all surfaces are exposed to the air (no overlapping pieces). You can use the racks from your oven or cooling racks as drying trays, or you can construct trays using stainless steel screening or cheesecloth stretched tightly over thin wood lath frames. Don't use aluminum, galvanized metal, copper, fiberglass, or vinyl for your trays because these can interact with the food to the detriment of its taste and nutrition.

Choose a spot away from dusty roads and yards, and cover the trays with tented cheesecloth to ward off insects and dust. Elevate the trays on blocks to allow good air circulation below the food. Beware of curious cats, free-range poultry, and investigative dogs when selecting a spot to place your trays of food. If you have raccoons or possums as night visitors, or any dewfall, bring your trays in at night.

You can dry fruits (including tomatoes) in direct sunlight, moving the trays periodically to keep them in the sun. However, vegetables will keep more of their color if set in a shady area.

You can purchase a solar dryer commercially or construct one at home. A solar dryer will dry food more effectively and more quickly than sun drying, although in less-than-optimal weather, you might have to opt for the oven or an electric dehydrator. You can find DIY plans for solar dehydrators online, in books, or through extension offices.

Oven drying is a practical way to get started with drying food because it doesn't depend on the weather or require a dedicated unit. To dry food in your oven, preheat the oven to a very low temperature (150 degrees Fahrenheit for vegetables or 170 degrees Fahrenheit for jerky). Put the prepared food on wire cooling racks (such as those used for cookies), making sure that all surfaces of each piece are exposed to air and that there is enough room between the pieces to allow heat to circulate. Put the racks in the warm oven. Rotate the drying racks carefully to ensure even dehydration. For oven drying, leave the door propped open a few inches to allow moisture to escape.

Oven drying does work, but if you plan to dry large quantities of food, consider investing in an electric dehydrator, which will save on energy costs and improve your end product because it can be difficult to maintain a low drying temperature in the oven. Oven dried foods are also darker, more brittle, and less flavorful than foods dried in a dehydrator. When you dry food outdoors or in an oven, it is sometimes recommended to pasteurize the food to kill any organisms that survived

Apricots, figs, apples, and dates are popular fruits for drying.

drying. Do this by preheating the oven to 200 degrees Fahrenheit and then placing the dried food on trays in the closed oven for fifteen minutes before storing.

Depending on which method you use and which food you're dehydrating, drying times will vary. Thicker slices take longer to dry than thin ones, of course. Solar drying takes longer (one to four days) than oven drying (twenty-four to forty-eight hours), whereas an electric dehydrator with a fan takes eight to twenty-four hours. Check for moisture content by bending a piece, be it green bean or beet slice. Most produce should be almost crisp, without discernible moisture, when fully dehydrated.

Place dried foods in a tightly closed large container. Stir or shake every day for a week. This will help equalize the moisture (those pieces that are too dry will absorb some of the moisture from those that are too wet). If the dried food still seems too moist, return it to the dryer/oven/sunshine for several more hours. Too much water content left in a few pieces can cause the whole batch to mold. Store your dried fruits and vegetables in airtight containers in a dry and dark place, where they will last for months to a year. To keep them longer, store in the freezer in airtight bags. You can store dried tomatoes in olive oil.

Rehydrating dried fruits and vegetables is easy: just cover them with boiling water and allow them to stand until the fruits/vegetables reabsorb the moisture. Once they are fully rehydrated, you can cook them. Another way to use dried vegetables is to eat them as snacks, add them to soups, or grind them into a powder and sprinkle them as you would a spice onto pasta, rice, bread, or soup. Sun-dried tomatoes, fruit chips, and fruit leathers also make unusual—and much appreciated—gifts for family and friends.

Fruit Leathers

Many fruits (and vegetables) can be pureed and then dried to create fruit leathers. Not only does the process result in a nutritious treat, but it also allows you to use slightly overripe (though unbruised) produce that doesn't make a first-class product when preserved by other methods.

To make fruit leathers, start by paring or peeling coarse-skinned fruits (such as apples) or washing thin-skinned fruits (such as berries) and then removing pits, seeds, and cores. Use a blender or food processor to puree the fruit, adding a small amount of water as needed to yield a very thick, yet pourable, puree. If the fruit is too juicy, strain it to make a thicker puree. You can also use home-preserved or store-bought canned or frozen fruit to make leathers (drain the fruit and puree as described).

Fruit leathers, such as these made with apples, are healthy snacks.

Add two teaspoons of lemon juice for every two cups of light-colored fruit to prevent darkening. If desired, up to half of the puree may be made of applesauce, which acts as an extender, decreases tartness, and makes the leather more pliable.

Use a baking sheet with sides to dry the leather. Place a sheet of freezer paper on the bottom of the pan, waxed side up, and then pour the puree onto it (a typical cookie sheet takes about two to two and a half cups of puree). Spread the puree evenly until it is no more than a quarter-inch thick across the bottom of the sheet. Place the cookie sheet in a dehydrator or oven set for 135 degrees Fahrenheit. Approximate drying times are six to eight hours in a dehydrator, up to eighteen hours in an oven, or one to two days in the sun. Turn the leather over onto another freezer paper-lined cookie sheet when you're able to peel it away from the first sheet of freezer paper; this will be about halfway through the estimated drying time.

Allow the leather to cool on a cake rack. Once cool, sprinkle it lightly with cornstarch to prevent it from sticking to itself and then roll it up.

Before and after: fresh basil leaves and dried basil.

Lavender is a popular herb for aromatherapy.

Drying Herbs

Depending on the herb, harvest may include one or more plant parts. In most cases, you harvest the leaves, but in some cases, you pick flowers, seeds, or roots as well. Often, you harvest blossoms with the leaves and mix them together. When stored properly, dried herbs retain their quality for two to three years. Although they will not spoil, they lose flavor after a while, so if you haven't used them up in that time, it may be time to add them to your compost pile and harvest a new bunch.

To maximize the flavor of herbs for cooking, harvest before the plants flower, early in the day, right after the morning dew dries. Cut annual herbs close to the ground; for perennials, never take more than a third of the plant at a cutting. After cutting, rinse the plants in cold running water and then pat them dry with a paper towel.

Hanging herbs upside-down is an easy way to dry them.

Although you can dry herbs in an oven or dehydrator, it is easy to dry them by hanging them upside-down in a cool, dark place. I use an old wooden drying rack, and I hang bunches of herbs in brown lunch bags to prevent dust from settling on them. Each bunch hangs with the top of the plant hanging down and the stems sticking up in the air; I attach the bag to the bunch with a rubber band and then tie the bag to the rack. This type of drying takes anywhere from a few days to a month, but unless you live in a particularly humid climate, there is little danger of the herbs molding as they dry. Of course, air drying requires no energy and retains most of the aromatic oils.

As with other dried foods, store dried herbs in a well-sealed jar in a cool, dark, and dry place. Although jars of herbs look pretty on a windowsill, direct sunlight fades them and reduces their potency.

Freezing Foods

The advent of the modern freezer changed our food system dramatically. Before the freezer's invention, food could be dried or canned for storage, but both methods took considerable time. The freezer allowed us to quickly preserve food by freezing. The main advantages of freezing are that it is simple and that it retains the original characteristics of the food (including texture and nutrients) more than other methods of preservation. One disadvantage, though, is that freezing any significant quantity of food takes a lot of room, ultimately requiring a stand-alone freezer, which is an expensive item and one you may not have room for.

Food should be stored in packages small enough to freeze quickly after it is put in the freezer; slow freezing causes large ice crystals to form and contributes to freezer burn. This also means that you should limit the quantity that you put into your freezer in any one day. Use adequate packaging, which also helps protect food from freezer burn. Although many people use rigid plastic or glass containers, I mainly use plastic freezer bags because they are convenient and inexpensive. They also tend to take up less space, and, if you drop them, they don't break! Make sure that you buy freezer bags, not just sandwich or storage bags, which are thinner and more prone to allowing freezer burn. The freezer bags with the zipper-type closure make it quite convenient to "burp" out all excess air, which is a crucial step in preventing freezer burn.

You can blanch corn on the cob and either freeze it whole or remove the kernels for freezing.

Freezing in Juice

Fruits often freeze best when packed in syrup or fruit juice or with a bit of sugar. Check in your canning book for options. Many light-colored fruits will darken in the freezer, but you can aid color retention by adding a tablespoon of lemon juice to your syrup.

For meat, fish, or corn on the cob, you can also use freezer paper taped up well with freezer tape to prevent air from getting to the food. If you are freezing liquids, such as fruits packed in syrup, juices, or sauces in rigid containers, leave at least one inch of empty space at the top; this headspace allows for expansion as the liquid freezes.

Whatever type of packaging you use, label all packages with the contents and the date on which they were frozen. Try to use a permanent marker so you can read the labels through frost if need be.

Canning Foods

The art of canning has been with us for hundreds of years, and an amazing and wonderful innovation it is. Canning heats food to a high temperature, thereby destroying enzymes and killing microorganisms that contribute to spoilage; then, as the heated jars cool, you create a vacuum so that air can't get to the food and new microorganisms can't grow (see sidebar on page 236).

Depending again on which fruit or vegetable you are canning, you will need either a water bather or a pressure cooker. The water-bath method involves boiling water in a water bather, which is

Water-Bath versus Pressure Canning

High-Acid Foods that Can Be Canned in a Boiling-Water Bath	Low-Acid Foods that Must Be Canned in a Pressure Canner
• Apples	• Artichokes
• Apricots	• Asparagus
• Berries	• Beans
• Cherries	• Beets
• Citrus fruits	• Cabbage
• Currants	• Carrots
• Figs	• Catsup**
• Grapes	• Corn
• Peaches	• Fish
• Pears	• Meats and Stews
• Pickles and Relish	• Okra
• Plums	• Peas
• Rhubarb	• Peppers
• Strawberries	• Salsa**
• Tomatoes*	• Sauces (BBQ, spaghetti, and the like)

* Add 1 tablespoon of lemon juice per quart.

** Some recipes have sufficient acid in the form of lemon juice, lime juice, or vinegar to be canned in a boiling-water bath.

essentially a large kettle with a wire rack to keep the jars from touching the bottom. This approach is suitable for highly acidic foods, such as pickles, relishes, chutneys made with vinegar, fruits, jams, jellies, fruit butters, and preserves. Place filled jars in the kettle and then pour in warm water until it covers the jars with about two inches of water. Water bathers made for this purpose will be big enough to give coverage and headroom for boiling water. After covering the jars with water, place the lid on and bring to a rolling boil. Process according to your canning chart. When the proper time has elapsed, remove the jars from the water with a jar lifter (available at most hardware and department stores). Place the jars away from drafts to cool. To make sure that the hot jars don't damage your countertop, set them on a folded dishtowel or on a cookie sheet.

Pressure canners are the workhorses of food canning, used season after season for thirty, forty, or even fifty years with little maintenance other than an occasional replacement of the seals and/or gauges. You must use a pressure canner with low-acid vegetables, meats, and fish to protect against botulism. The typical size holds seven quarts, or fourteen pints, at a time. While pressure canners are a bit pricey, you will need one if you plan on canning regularly. Do some comparative shopping, buy a reputable brand, and think of it as an investment that will continue to pay off year after year.

The water-bath method processes the jars in boiling water.

When looking at pressure canners, you'll see that there are two basic types: those with a gauge and those with a petcock. There are proponents of both types. Those who have used one with a gauge find it convenient to be able to read the exact amount of pressure at which the canner is operating. One drawback with this style is that you must have your gauge calibrated annually to keep it accurate because if your food does not reach a high enough temperature to kill harmful bacteria, it will be unsafe to eat. Likewise, if too much pressure builds up, you will have a dangerous situation on your hands. If your safety valve blows, you'll be in danger of not only steam burns but also of the glass jars' breaking and shooting pieces of glass and food up out of the safety valve. Most extension offices will test and calibrate your gauge for you.

The alternative is a pressure canner with a petcock, which is a weight that sits on top of a steam vent in the center of the lid. Typically, a petcock for a pressure canner has three different sized holes, each labeled with a certain weight: 5 pounds, 10 pounds, and 15 pounds. You choose the appropriate weight depending on what is needed for the specific food. For example, your canning instructions might tell you that "for quarts, use 10 pounds of pressure for 25 minutes."

A drawback of this style is that you must be vigilant in listening for the jiggling of the petcock as it lets out excess steam to be sure that it is operating at the correct temperature. Ideally, jiggling four to five times a minute indicates that the temperature is right. If it jiggles only once or twice per minute, the temperature is too low. The real beauty of this simpler model is that if you let it jiggle continuously for short times, nothing bad will happen; however, if you let it jiggle continuously for an hour, it will likely boil dry, thus releasing the pressure too quickly and causing the jars to break.

A petcock model also has a safety valve, but unless the vent hole becomes plugged, you will never need to activate it. As a precaution, though, always rinse out your canner before loading and look or blow through the vent hole to make sure that there is no foreign debris that could plug it.

Be aware that petcocks sit loosely on the vent, so they can be easy to lose. One veteran canner has stored her petcock in the same place in her spice cupboard for thirty-one years, but many people end up scrambling to order new ones when harvest season arrives.

Pears are suitable for canning by the water-bath method.

One caveat for petcock canners—it's tempting to "hurry" the time required for the pressure to decrease by jiggling the petcock, but if you do it too much, releasing the pressure too swiftly will cause liquid to suck out of the jars, possibly preventing a good seal.

Something else to consider when buying a pressure canner: All lids have seals, but different models have different means of fastening the lids on. Some use clamps, which must be screwed down and then released once the pressure has completely dissipated after the jars have been processed. Others twist on and have a safety lock in the handle to prevent lid removal until the pressure has gone completely down and it is safe to open. Some types are easier to use than others.

Each year, when I first use my pressure canner, I test the petcock and the gauge before I begin canning. To test a pressure canner, fill the canner with water, but no jars, and then seal it but leave the petcock open. As soon as the water begins to boil, steam will escape through the petcock. If steam isn't escaping, there's a plug in the petcock and you need to clean it. If steam is escaping, close the petcock and watch the gauge. It should begin to climb quickly. If it isn't climbing, you need to replace it.

Before processing jars, you can pack jars using either the hot-pack or raw-pack method. For the hot-pack method, prepare food (as in the case of sauces and spreads) or briefly boil it in water, syrup, or juice and pack while still hot into preheated jars. Then add the cooking liquid or boiling water. For the raw-pack method, pack raw food in jars and cover with boiling water, syrup, or juice. With either method, leave from one half to one inch of headspace in the jar to allow for expansion when hot.

The hot-pack method has several advantages over raw packing. Heated fruits are easier to pack into jars because they are softer, thus more fruit can be put into each jar. Therefore, fewer jars are used for the same amount of food, so the processing time is usually shorter. Hot-packing better protects the color

Canning Jars

Use glass jars specially made for home canning. Canning jars come in narrow-mouth and wide-mouth styles. The narrow-mouth style works fine for things like jellies, jams, and relishes, but I prefer the wide-mouth jars for everything because I simply find them easier to work with and easier to clean.

Glass canning jars come with metal lids and rings to hold the lids on during processing.

of fruits such as apples and peaches. Hot-packing does, however, require the extra step of cooking or boiling, and thus is more time consuming.

When you finish processing the food, and the jars have cooled off, clean them with a damp cloth and check to make sure that the lids have sealed properly. Manufacturers design canning lids so that the raised button in the center of the lid flattens out if you've established a proper vacuum. Most cooks remove the jar rings right away, wash and dry them, and save them for next time because any moisture trapped underneath them may generate rust, not only ruining the ring but also sometimes rendering the jar nearly impossible to open.

Homemade Fruit Juice

You can prepare fruit juice for drinking, for cooking, or to use as the critical ingredient in jellies and other spreads. First, wash the fruit and drain it well. For large, round fruits such as apples or pears, quarter them, core them, and remove the stems; for berries and grapes, remove the stems. Crush the fruit in the kettle, and if the fruit doesn't have a good deal of natural juice, add about half a cup of water to keep the fruit from sticking to the pan. Begin cooking over low heat, stirring frequently. As you draw natural moisture from the fruit, increase the heat to medium. Some dry fruits (such as apples) may require more water, but add it in small amounts (a quarter-cup at a time) so you don't overdilute. Continue to cook the fruit until it loses its color.

After you've cooked down the fruit, collect the liquid in a pot. Do this by pouring the mash and juice into a wet jelly bag or into a colander or strainer lined with wet cheesecloth (or unbleached, undyed muslin kitchen towels kept for this purpose). To get the clearest juice, do not squeeze the bag, but let it drain naturally (overnight works well).

Canning Tips

As you will find in canning recipes and instructions, the amount of time needed to process the jars increases with elevation. Add five minutes for every 2,500 feet of altitude above 1,000 feet above sea level. For example, if a recipe calls for ten minutes' processing time, that time is appropriate for locations up to 1,000 feet above sea level. Between 1,000 feet and 3,500 feet above sea level, process for fifteen minutes, and for 3,500 to 6,000 feet above sea level, process for twenty minutes.

All jars are not created equal. Although it's tempting to stockpile those mayonnaise jars to use when it's time to can green beans, don't bother. I'm sorry to tell you this, but those jars are not meant to stand up to the heat and pressure needed to safely preserve food. Antique jars, too, are not a good choice; although they may be strong, they may not stand up to the pressure. Discard any jars that are chipped on the rim because all it takes is one small nick to prevent the lid from forming a seal. Reusing the lids is also not recommended, even though on the surface it appears to be a nifty-thrifty plan. Because the rubber on the seal has already been used once, the risk of an incomplete seal rises. However, you can reuse jars as well as the screw rings that hold down the sealing lids, provided that they are not rusted or misshapen.

Fruit Cocktail

1 ¼ pounds slightly underripe seedless grapes

¼ cup lemon juice mixed into 4 cups water

3 pounds peaches, ripe but firm

3 pounds pears

10-ounce jar maraschino cherries

3 cups sugar

4 cups water

Stem and wash grapes. Soak grapes in lemon juice and water solution. Dip peaches, a few at a time, into separate pot of boiling water for 1–1½ minutes to loosen skins. Dip peaches immediately in cold water and slip off skins. Cut in half, remove pits, cut into ½-inch cubes, and add to solution with grapes. Peel, halve, and core pears. Cut into ½-inch cubes, and combine with grapes and peaches. Place mixed-fruit pieces in colander to drain excess liquid. Combine sugar and water in saucepan and bring to boil to form syrup. Add ½ cup of hot syrup to each jar. Add a few cherries and gently fill each jar with mixed fruit and more hot syrup, leaving ½-inch headspace. Adjust lids and process.

Yield: About 6 pints.

Pickling

Harvest cucumbers no more than twenty-four hours before pickling, if at all possible, or you may end up with hollow pickles. Some varieties of cucumbers are not suitable for pickling because they make soggy pickles. Pickling varieties have a nubbier texture than eating varieties.

When pickling foods, whole fresh herbs and fresh spices provide the best flavor. Although you can leave some herbs, such as dill or chives, in the jars when the pickles are canned, you should remove most of the herbs and spices when the pickles are ready to go in the jars. The easiest way to do this is to place herbs and spices in a sterile cheesecloth bag to infuse flavor during the pickling process and then just remove the bag before canning.

Mix vinegar and water, salt and water, or a combination of vinegar, salt, and water to make a brine solution for the pickles. Temper the brine solution to taste with a sweetener. Use a good-quality, aged vinegar product and salt labeled "pickling salt." Never use brown sugar as a sweetener unless the recipe you are following specifies it because it tends to darken the pickles to an unappealing color. You can readily interchange honey for white granulated sugar for sweetening, depending on your preference.

Although we usually think in terms of cucumber pickles, all kinds of produce can be pickled or made into relishes and chutneys. Experiment with pickling your favorite fruits and veggies—from asparagus tips to watermelon rind. And don't forget that you can also pickle other foods, such as eggs, fish, and meat.

The water that you use for pickles is crucial to pickle production. When you prepare pickles with water that is high in mineral content, or with soft water from a commercial water softener, the pickles may turn out mushy and unappetizing. In these cases, try pickling with bottled distilled water.

Pickled Cucumbers

> 6 lb. ¼-inch-thick pickling cucumbers, cleaned and sliced
>
> 4 cups onions, chopped
>
> 4 cups ice cubes
>
> 4 cups vinegar
>
> 4 cups sugar
>
> 2 Tbsp. celery seed
>
> 2 Tbsp. mustard seed
>
> 1 Tbsp. ground turmeric
>
> ½ cup canning salt

Combine cucumbers, onions, and salt in large bowl and mix well. Cover with 2 inches of ice cubes and refrigerate for four hours. Combine remaining ingredients in large kettle and boil for 10 minutes. Drain the liquid from the cucumber/onion mix and add the vegetables to the kettle. Bring back to boil for 2 minutes. Pack pint jars, leaving ½-inch headspace. Adjust lids and process in a water-bath canner for 10 minutes. Yield: 8 pints.

Tomato Sauces and Salsas

The best tomatoes for sauces and salsas are paste, or Italian, types. These create a thicker sauce or a salsa that sticks to a chip. If your garden is overrun by slicing tomatoes, you can use them with paste tomatoes (shoot for 30 percent slicing tomatoes to 70 percent paste tomatoes) for a medium-thick sauce or salsa, or you can add tomato paste to the recipe to thicken it.

Botulism

Clostridium botulinum is a naturally occurring, soil-borne bacteria that causes the potentially fatal disease known as *botulism*. Unlike most food spoilage bacteria, *C. botulinum* thrives in the anaerobic (or oxygen-free) environment of canned food; both commercially canned and home-canned foods can carry the organism. As it reproduces in the canned food, it creates a deadly nerve toxin.

The classic symptoms of botulism include double or blurred vision, drooping eyelids, slurred speech, difficulty swallowing, dry mouth, and muscle weakness; these are all signs of the muscle paralysis that this bacterial toxin causes. If untreated, these symptoms may progress to cause full paralysis of the arms, legs, trunk, and respiratory muscles.

In food-borne cases of botulism, symptoms generally begin eighteen to thirty-six hours after eating a contaminated food, but they can occur as early as six hours or as late as ten days after exposure.

Food-borne botulism is most often associated with home-canned foods that have a low acid content, such as asparagus, green beans, beets, and corn. However, there have also been reported outbreaks of botulism from more unusual sources, such as chopped garlic in oil, chile peppers, tomatoes, or improperly handled baked potatoes wrapped in aluminum foil.

Home canners should follow strict hygiene procedures to reduce the risk of contamination and must can low-acid foods with a pressure canner instead of a water bath. The temperature in a water bath is not high enough to guarantee killing all *C. botulinum* organisms.

For recipes that call for peeled tomatoes, dip the tomatoes in boiling water until the skins begins to split (usually less than a minute) and then immediately place the tomatoes in cold water. After this double dip, you can easily peel off the skins.

Don't have the equipment or time to deal with canning your sauce or salsa? Try freezing it. Make your recipe. Once cooled, fill appropriate freezer containers and store in the freezer. The sauce or salsa will be slightly watery when thawed, but heating for a few minutes on the stovetop will cook off the water. For salsa, allow it to cool again before serving.

When jars are properly canned, the lids will retain a tight seal when you remove the rings.

Spaghetti Sauce

    30 lb. tomatoes

    1 cup onions, chopped

    5 cloves garlic, minced

    1 cup celery or green peppers, chopped

    1 lb. fresh mushrooms, sliced (optional)

    ¼ cup vegetable oil

    2 Tbsp. oregano

    4 Tbsp. parsley, minced

    2 tsp. ground black pepper

    4½ teaspoons salt

    ¼ cup brown sugar

Wash tomatoes and dip in pot of boiling water for 30–60 seconds or until skins split. Dip immediately in cold water and slip off skins. Remove cores and quarter tomatoes. Simmer for 20 minutes, uncovered, in large saucepan. Put through food mill or sieve. Sauté onions, garlic, celery or peppers, and mushrooms (if desired) in vegetable oil until tender.

Add sautéed vegetables to tomatoes and mix in remainder of spices, salt, and sugar. Bring to a boil. As soon as mixture boils, reduce heat and simmer, uncovered, until thick enough for serving (volume should reduce by nearly half). Stir frequently to avoid burning. Fill jars, leaving 1-inch headspace. Adjust lids and process. Yield: About 9 pints.

Catsup

24 lb. ripe tomatoes

3 cups onions, chopped

1 tsp. ground red pepper (cayenne)

4 Tbsp. pumpkin pie spice

3 cups apple cider vinegar (5 percent)

1½ cups sugar

¼ cup salt

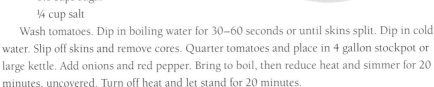

Wash tomatoes. Dip in boiling water for 30–60 seconds or until skins split. Dip in cold water. Slip off skins and remove cores. Quarter tomatoes and place in 4 gallon stockpot or large kettle. Add onions and red pepper. Bring to boil, then reduce heat and simmer for 20 minutes, uncovered. Turn off heat and let stand for 20 minutes.

Add spices and vinegar to tomato mixture. Turn heat on and boil mixture for about 30 minutes. Put boiled mixture through food mill or sieve. Return to pot. Add sugar and salt; return to gentle boil. Cook, stirring frequently, until volume is reduced by half or until mixture rounds up on spoon without separating. Fill pint jars, leaving ⅛-inch headspace. Adjust lids and process. Yield: 6 to 7 pints.

Tomato Salsa (Mild)

4 cups tomatoes, peeled, cored, and chopped

2 cups long green chile pepppers, mild, seeded and
    chopped

1 cup onions, chopped

4 cloves garlic, finely chopped

2 cups vinegar

2 tsp. ground cumin

2 tsp. fresh cilantro (or 2 tsp. dried cilantro)

1½ tsp. salt

Combine all ingredients in large saucepan and bring to boil, stirring frequently. Reduce heat and simmer for 20 minutes, stirring occasionally. Ladle hot salsa into pint jars, leaving ½-inch headspace. Adjust lids and process in water-bath canner for 15 minutes. Yield: 4 pints.

Jellies, Jams, and Other Sweet Things

When fruit is abundant, use it to create sweet spreads such as jelly, jam, and fruit butter. These sweets are preserved with added sugar, and many require the addition of pectin, which is a jelling agent, or acid. Pectin is a naturally occurring jelling agent found in all fruits, but some have an abundance. Depending on the type of fruit and its ripeness, pectin may not be available in sufficient quantity to produce good spreads.

Apple Butter

8 lb. apples (use Jonathan,
    Winesap, Golden Delicious, or
    Macintosh for good results)
2 cups apple cider
2 cups vinegar
2¼ cups white sugar
2¼ cups packed brown sugar
2 Tbsp. ground cinnamon
1 Tbsp. ground cloves

Wash, remove stems from, quarter, and core fruit. Cook slowly in cider and vinegar until soft. Press fruit through a colander, food mill, or strainer. Cook fruit pulp with sugar and spices, stirring frequently. To test for doneness, remove a spoonful of apple butter and hold it away from steam for 2 minutes. You've finished cooking when the butter remains mounded on the spoon. Another test is to spoon a small quantity onto a plate. When a rim of liquid does not separate around the edge of the butter, it is fully cooked. Pour hot apple butter into sterile half-pint or pint jars, leaving ½-inch headspace. Adjust lids and process. Yield: About 8 to 9 pints.

A peanut-butter-and-jelly favorite—grape.

Homemade
canned
apricot
preserves.

Pear-Apple Jam

1 cup apples, peeled, cored, and finely chopped
2 cups pears, peeled, cored, and finely chopped
½ tsp. ground cinnamon
6½ cups sugar
¼ cup bottled lemon juice
½ cup pectin

Crush apples and pears in large saucepan and stir in cinnamon. Thoroughly mix sugar
and lemon juice with fruits. Bring to boil over high heat, stirring constantly. Immediately stir
in pectin. Bring to full rolling boil and boil hard for one minute, stirring constantly. Remove
from heat, quickly skim off foam, and fill sterile jars, leaving ¼-inch headspace. Adjust lids
and process. Yield: About 7 to 8 half-pints.

To test fruit to find out if it needs added pectin, mix 1 tablespoon of cooked fruit or fruit juice
with 1 tablespoon of rubbing alcohol. When the fruit has sufficient pectin, the mixture will
coagulate into a clump. If it doesn't clump into a single blob, add pectin. Throw test samples away
because rubbing alcohol is poisonous.

Preparing homemade pectin is easy if you have access to plenty of sour apples (underripe eating
apples or crab apples). Wash and quarter 10 pounds of sour apples and remove the stems but not
the skins, cores, or seeds. In a large kettle, cover the apples with cold water and bring slowly to a
boil over medium heat. Simmer at a low boil for 30–45 minutes until the fruit is quite soft. Drain

Jelling Properties of Fruit

Fruits that Normally Have Sufficient Pectin and Acid to Gel	Fruits that Benefit from Additional Pectin or Acid	Fruits that Will Require Additional Pectin or Acid
• Apples (green or crab)	• Apples (ripe)	• Apricots
• Blackberries (sour)	• Blackberries (ripe)	• Cherries (sweet)
• Cranberries	• Blueberries	• Grapes (western)
• Grapes (eastern)	• Cherries (sour)	• Pears
• Lemons	• Grape juice	• Peaches
• Limes	• Oranges	• Raspberries
• Plums		• Strawberries
• Quinces		

the fruit as you would for making juice or squeeze the pulp to get all of the juice out for pectin production. Store your fresh pectin in the refrigerator if you plan to use it within a few days; otherwise, freeze it.

Quick recipes using commercially available pectin allow you to make spreads more quickly than the old method of cooking down the fruit in an open kettle. These recipes require far more sugar, however. Most fruits are acidic, but they may not have sufficient acid for proper jelling to occur, which is why many recipes call for the addition of lemon juice or vinegar.

I know some cooks who have been making fruit spreads since before I was born and don't need to use a candy thermometer. But for those of us who lack these skills, a candy thermometer is an inexpensive but very important investment for producing spreads with the right consistency. The mixture of sugar and fruit or juice is usually cooked sufficiently when it reaches 220 degrees Fahrenheit for spreads other than butters (subtract 2 degrees for every 1,000 feet above sea level) and 300 degrees Fahrenheit for fruit butters.

Fruits can easily scorch when you are making spreads, so you must stay by the pot. Stir frequently, making sure to scrape the bottom of the pan as you do, and watch your candy thermometer. You can use oven cooking techniques to reduce the likelihood of scorching, but cooking in the oven increases the time—and the effort— needed to prepare the spread because you have to pull it out of the oven to stir it periodically.

Preserving the Harvest: Dairy and Meat

Dairy products and meat from your own animals that have had access to plenty of sunshine, fresh air, good pasture, and clean water can't be beat. They provide not only great flavor but also more micronutrients (such as omega-3 fatty acids, conjugated linoleic acid, and vitamin E) than their feed-lot or factory-farm counterparts.

Dairy Products

With a dairy cow or dairy goats, you can have your own fresh milk and produce your own butter, cheese, ice cream, yogurt, and myriad other milk-based products.

Milking Your Animals

Getting milk from the animal and into your kitchen is an age-old process, and although commercial-scale operations require the use of milking machines to get the whole herd milked twice a day, for a family cow or a couple of goats, you can do it the old-fashioned way quite easily unless you're plagued by arthritis. If you can't or if you don't want to hand milk, purchase a small automatic milker from a dairy supply company. If you decide to go this route, be sure to do your research to find the one that fits your operation the best. Milking is typically done twice a day, as close to twelve-hour intervals as possible, although many hobby farmers are turning on to the beauties of once a day (OAD) milking.

Be sure to start the milking process with clean, dry udders.

Hand milking isn't rocket science, and, just as with any skill, you can get the hang of it with a bit of practice. It's ideal if you can visit a farm that milks cows or goats by hand and observe their process. Watching a live demonstration with a running tutorial is a great help, and the farmer might even let you have a bit of "hands-on" practice as well. Alternately, in this Internet age, you can search online for videos on how to milk a cow by hand, and you'll find all sorts of help. Use your own common sense when evaluating a farmer's method and deciding on what practices you want to emulate.

Cow or goat, the milking process is virtually the same, with the exception that goats have two teats and cows have four. To simplify, I'll proceed with the explanation as if you were milking a cow. To begin, put the animal in a stanchion (a head-holding device) or restrain her with a halter or collar clipped to a rope that is securely tied (a horse's lead rope works great for this). Many farmers offer a bit of feed at this point as an incentive to come in for milking as well as some hay to keep the animal occupied (and thus standing more quietly) while she's being milked. After securing your cow, give her side and udder a quick brush-over with your hand to get rid of any dust or debris that could fall off into your milk bucket once you start milking. During the winter in snowy regions, brushing off the snow and then at least partially drying her back and side above where you'll be sitting will help prevent melted snow from dripping onto you or into your bucket.

Next, wash her udder and teats with warm soapy water and a soft rag. Unless you're one of the few who have hot running water in your barn, just bring some hot water from the house with you in your milk bucket, dump it in a coffee can or small bucket dedicated to udder washing, add a couple drops of dish soap (leave a bottle on the shelf near your milking station), and you're ready to wash. Squeeze your rag out well, as you don't want to get the udder so wet that residual water drips into your bucket while you're milking. Unless ol' Bessie laid down in a manure pile, her udder will be pretty clean anyway, but rubbing it over with the cloth rids the milk bag of dust, dander, and debris and helps her start "letting down" her milk, as does washing the teats. Rinse your rag, hang it to dry, toss out your wash water, and you're ready to roll.

You'll need a stool. This can be as simple as the old "T" stool (two short lengths of two-by-four nailed perpendicular to each other), a plastic step stool from your hardware store, or just about anything that is lightweight enough to position with one hand and the right height to let you sit beside your cow and easily reach her teats.

Sit beside her on your stool; cows are usually trained to be milked from the right side. Tuck the bucket between your knees, tilted a bit toward the cow so you can get milk into it without sitting it directly under her (which is an open invitation for dirt, debris, and being kicked over). Positioning it this way will also enable you to shelter the pail from most foreign debris as you milk as well as make it easier to grab if your animal moves.

Grasp one teat in one hand between your thumb and index finger in a sort of straight-fingered clamping motion right at the base of the udder where the teat attaches. This is important, because if you just grab a teat and squeeze, the milk that is in the teat will partially go back up into the udder, and you'll only get a few drops. Once clamped, roll your fingers around the teat and squeeze. Although a natural downward motion seems to happen as you clamp, squeeze, release, and repeat, the pulling itself doesn't extract the milk. Rather, as you empty the milk in the teat and release for a second before clamping again, gravity brings in another teatful.

Once you get going, the milk will come out in quick, even streams.

Discard the first several squirts of milk to ensure a clean stream, and collect the rest in the clean milk bucket you brought with you from the house. It is most efficient to milk two teats at once, and you will need to fully strip them of milk. This stripping is important because the cow's milk supply responds to demand. Not fully emptying the udder signals the animal's body to make less milk next time around, and, if done often enough, will cause the animal to "dry up" before she normally would, leaving you milkless.

First, milk her out until your streams dwindle to trickles, at which point it's time to "strip" the teats. Do this using only your clamped thumb and index finger, starting as high up on the udder as you comfortably can and sliding downward to the end of the teat. Repeat. The creamiest milk is in these strippings. You can be confident that you've emptied the udder sufficiently when this stripping action yields little or no milk. Be sure and strip all four teats.

Sometimes, your cow will move around. The time-honored command here is "Saw," which means "stand-still-you-rascal-I'm-trying-to-milk-here." It can also be helpful to plant your head in her flank. This inhibits the cow's free movement and will likely make her subside. As an added bonus, it enables you to feel when she's about to shift her weight, move her foot, or, heaven forbid, decide to kick.

I know that a kicking cow sounds dangerous, but it rarely is. Unlike a horse, cows don't kick backward. Rather, they fling their back legs forward and out. If your animal decides to do that, grab

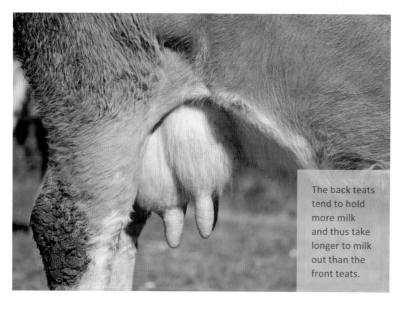

The back teats tend to hold more milk and thus take longer to milk out than the front teats.

your bucket with your right hand (if you're milking from the cow's right side), swing it out of harm's way, block/deflect her kick with your left forearm, and say, "Saw!" sternly. Some milkers use stanchions that include a blocking bar inserted in front of the cow's back legs to prevent kicking, but that is not normally needed. If your cow tends to dance around a bit, consider bringing a second empty bucket to the barn with you and switch out about halfway through—that way, if the cow does manage to knock your bucket over, you don't lose the whole milking.

You will find that the udder's back quarters usually hold more milk than the front quarters, and that many animals prefer to be milked from only one side. Because of this, consider milking both back teats at the same time, and milk them first while your cow is standing still and occupied with her feed; they tend to be more difficult to reach. Empty them until they need to be stripped and then milk out the front quarters the same way. Once you've milked out all four teats, go back to strip them.

If you opt for milking from both sides, taking front and back teats on the same side together, you'll find that the front teats finish before the back teats, so ultimately you will end up milking the two back teats together to finish off. That said, outside of completely stripping the udder, there is no one right way to "get-er done."

Next, to help prevent bacteria from entering the teat while the sphincter muscle at the opening is relaxed (it can take a couple of hours to tighten back up after a milking), dip each teat in iodine solution, a commercial teat-dip product, or a homeopathic concoction of antibacterial essential oils. In the winter, you may want to also apply a thin coat of udder balm to the teats themselves to keep them from becoming dry and cracked.

If you're new to the milking process, hang in there! It's much more organic and commonsense than it sounds on paper. After all, people have been doing it since time immemorial, and you will be no exception, even if your hands get very tired before you're finished and your cow/goat becomes impatient. Fork her some more hay and carry on.

You don't have to do it perfectly the first time. The milk will not dry up overnight. Your hands will get stronger, and, before you know it, you'll be an old hand, teaching your children or grandchildren how to milk. Be encouraged and know that there is something comforting and satisfying about the contented munching of a cow, the warmth of her side against your shoulder or head, the singing of milk into the bucket, and the sounds of the farm all around. Rare moments, these, and sweet rewards for our efforts.

Transitioning to Once-a-Day Milking

In this busy day and age, needing to milk twice a day can be a real time challenge. Good news! Why not milk just once a day? Farmers from New Zealand to California are doing it with complete success. The benefits are readily obvious: one less session of washing all of the milking equipment and traipsing down to the barn, washing the cow, milking, and then straining and storing the milk adds up to time and labor saved. If you decide to switch to OAD milking, the process will take a week or more to accomplish, but it's not complicated. Just choose your optimal milking time—morning, noon, or evening. Let's say morning, for the sake of example.

If you are currently be milking at 8:00 a.m. and 6:00 p.m. and want to eliminate the evening milking session, begin transitioning to OAD by moving the evening milking to 3:30 or 4:00 in the afternoon. This earlier milking will yield less milk than your usual evening milking, and your cow's udder will be fuller the next morning, but not enough to cause her discomfort. Maintain this new schedule until the udder is not overly full in the morning milking. Then, subtract two or three more hours from the second milking, repeating the entire process perhaps a total of three times, waiting each time for the udder to return to normal fullness.

By this time, the cow will have regulated her supply of milk to meet your demand. This does not mean that she gives half the amount she did when you were milking twice a day, but she will have dropped somewhat in production. However, unless you can use multiple gallons of milk a day in some way, a lower volume may work just fine for you.

Another option for OAD is "share milking"—leaving the calf with the cow while still milking. At first, you will have to milk twice a day because a newborn calf cannot handle the amount of milk that a dairy cow can produce—depending on the breed, a cow can produce more than five gallons a day! One Jersey cow owner who decided to share milk said, "The benefits are tremendous. I get all the milk I need, both mama and baby are supremely happy and healthy, I don't have to deal with

gallons of milk every day. As an added benefit, I know that if I have to miss my usual milking, they will not mind at all."

Since your OAD milking is taking the excess milk that the calf cannot handle, you ward off milk scours, you save yourself the trouble of bottle feeding, and you give the calf the benefit of learning to graze earlier from its mother than their pen-raised counterparts. Nursing calves are healthier because they can nurse as needed rather than overfill their bellies twice a day. Nursing also fends off mastitis in the cow, because the calf suckles at will, and milk will not be sitting in her udder for hours at a time. The cow's

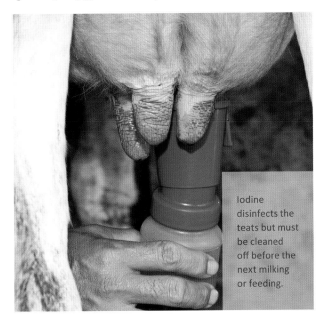

Iodine disinfects the teats but must be cleaned off before the next milking or feeding.

Grafting on a New Calf

Getting a cow to adopt a calf can be easy or challenging, depending on your cow. On the Bartell farm in western Montana, Gordon, who raised Black Angus cattle and always had a couple of milk cows, had many occasions to assist in this process over his forty years of ranching. Typically, these situations arose from trying to pair up a beef cow who had lost her calf with a different newborn needing a mama. For a decade or more, he also had a Holstein milk cow that gave enough milk for three calves and the kitchen, so there were years when the big old black-and-white cow was followed by her own black-and-white baby and a couple of Black Angus siblings.

Cows know their own offspring largely through their sense of smell, so Gordon's methods involved dousing the mother cow's nose and the adoptee from head to tail with a liberal dose of a strong scent (one year, he used an aftershave called "Hai Karate") so that the cow and the calf smelled the same. With a more docile cow, Gordon would observe the interaction between calf and cow. With a wilder cow, he would tie her head, leaving her enough slack so she could reach around and smell the baby. Some cows still kicked the calf away, which meant that mealtimes had to be supervised for several days, but most cows did go ahead and adopt a calf, to their mutual satisfaction. Milk cows are more usually much more docile by nature and used to being handled than beef cattle, so you usually don't have to work as hard to graft a calf on. The aforementioned Holstein was so adaptable that she'd often let other calves stop in for an extra meal even if the calves had perfectly good mothers of their own.

Gordon's daughter grew up milking cows twice a day from the age of six. Now a hobby farmer and cow owner herself who practiced share-milking, she fell so much in love with the concept of milking only when she needed milk that when her Jersey's first calf, Sooner, was pushing six months old and ready to be weaned, she bought a three-day-old calf from a local dairy (a good source for baby calves because they usually sell all bull calves and a few heifers every year).

One morning, Sooner was sent to a different pasture, which was the usual procedure for separating him from his mother when the farmers needed some milk. That evening, good old Rosie met her new baby boy. To help the process along, Rosie's owner used her dad's method, substituting blackstrap molasses for aftershave trickling it down the calf's head, back, sides, and tail and then turning him in with the cow. Rosie took one sniff and started licking, while the owner helped "Bucky" get latched on.

The owner said, "Rosie didn't kick once, and when it was time for her to exit the barn, she didn't want to leave him behind. In fact, she spent the night camped out on the other side of the door." "Bucky nursed the next morning, and she licked him all over again, without the molasses. It truly was the easiest grafting on I've ever seen, and I've seen and helped with quite a few! It was as if Rosie had given birth to Bucky herself. And to make it even sweeter, our old Welsh pony, Ginger, also adopted Bucky, standing over him while he slept and Rosie grazed, following him like a nervous aunt when he ran and played. They were such a happy threesome."

Also, because Sooner was so used to being separated from Rosie when the family needed milk, he hardly missed her at all. At the age of six months, his rumen was fully developed, so milk was more of a habit than necessary nutrition. Likewise, Rosie was so occupied with Bucky that she didn't miss Sooner, either. The owner relates that Rosie was bred about three months later, and Bucky was weaned at six months old. When it was all said and done, before they dried her up, Rosie had been giving the family milk, butter, and cheese for a year and a half, and she was milked only when they needed milk—two or three times a week as opposed to twice a day. In adddition, she'd raised two steers, which the family did eventually add to their freezer in due course.

milk production will not drop as it does with OAD because the calf and you are taking all of the milk that she can make.

As the calf grows and is able to take more of the milk, you will eventually want to shut the calf away from the mother at night whenever you want to get some milk—a three- or four-month-old calf can probably handle most, if not all, of the cow's milk. This is a beautiful stage, because it means that you can let the calf "milk" the cow until you need another gallon in your fridge—say every two or three days. In this way,

"Share milking" benefits the cow, the calf, and the farmers.

you can have all of the advantages of a milk cow with very little time investment. If you become interested in share milking or milking OAD, there are plenty of helpful websites and support forums to facilitate your journey. The calf you're sharing with can go on to join your herd or, if he's a steer, to eventually be weaned and move on to his next stage of life. When the first calf is weaned, if your cow is docile and amenable, you can "graft" another newborn calf on and start the cycle again, as the aforementioned Jersey cow owner did (see sidebar on page 248).

Processing Your Milk

Once you have milked, it's time to strain and either refrigerate or process the milk. You'll need a container to store the milk in. Glass jars are ideal because they won't harbor bacteria that can spoil your milk, they chill quickly, and they clean up easily; the only downside to glass is the potential for breakage. Stainless steel is also a good choice. Aluminum is not. Plastic is all right, but it does have

tiny pores that can hide bacteria. If you want to reuse store-bought plastic milk jugs, know that it is virtually impossible to remove all traces of previous milk, and this residue of "riper" milk will cause your new milk to sour quicker. Yes, you can bleach the jug, but it does not remove the milk film (or *milkstone*, as the film of calcium is sometimes called), and this film contributes to the harboring of bacteria that will shorten the refrigerator life of your milk.

You'll also need a strainer and filters. Way back when, people used sterilized muslin, layers of cheesecloth, or some other fabric to strain any foreign particles out of milk. Nowadays, you can purchase commercial filter discs that fit into the mouth of the strainer, which is basically a large stainless-steel funnel found in various types and sizes. A word on strainers here: Unless you're running a large operation and handling considerable volume that you strain

into large-mouthed cans, go with a strainer that fits into the mouth of a gallon jar (like a pickle jar). That way, you can set the strainer directly on the jar rim while you pour milk into it. Measure your jars of choice and find a strainer that will fit in their openings.

Prepare your strainer by washing it in hot soapy water and rinsing with very hot water. Fit the strainer disc into the strainer according to manufacturers' directions. Next, set your strainer on your clean jar and pour the fresh warm milk into the top. A word of caution here: Don't just dump the whole bucketful into your strainer at once unless you know that your milk jar will hold it all. Take it from one who's "been there, done that"—it is difficult to transfer a strainer full of milk to the next jar or back to the bucket without making a mess.

After you've strained the milk, put a lid on the jar and refrigerate it. Wash your milk bucket and strainer so they are ready for next time.

Cleanliness

Just some quick advice here for keeping things clean during milking. First, the outside of your bucket will get dirty. With milk splatters, mud, and manure from the floor or a cow's foot—even though you take precautions, such as putting fresh hay, sawdust, or straw under your milking area—buckets can start looking pretty gross. It may go without saying, but I'll say it anyway—what's on the outside of the bucket should not come in contact with anything that goes in your mouth. The kitchen sink, used to wash food and dishes, is therefore not the place to deal with the outside of a milk bucket. This is probably the biggest thing to remember, as well as that whatever you use to scrub the outside of the bucket—brush, cloth, or sponge—should never be used for the inside of the bucket or anything else you'd eat from. We're talking potential *E. coli* contamination here. You can figure out what works best for you. A utility sink works well, for example. Here are the basics in six easy steps:

- Make some hot soapy water with a bit of bleach in a bucket or pan.
- Set your milk bucket down in the water (if your basin is too small, set it in at an angle) and, with a small brush, such as a fingernail brush (handy for the "stuck-on" parts), and/or a cloth, wet down the entire outside of the bucket.
- Allow to soak for five or so minutes and then scrub clean.
- Rinse with clean, hot water.
- Clean your brush and set aside for next time. Throw your cloth into the washer.
- Wash the inside of the milk bucket with hot soapy bleach water, rinse, and dry—you're ready to go! (Note: Do not use the same towel to dry the outside and inside of the bucket unless you dry the inside first. You can then throw the towel(s) in the wash with the cleaning rag.)

Not only is cleanliness in the kitchen important, cleanliness in the barn is also critical from start to finish in dairying, especially if you're not share-milking. As previously mentioned, milk remains in the udder for much longer than it does if the calf (or kids) is suckling multiple times throughout the day, which means a higher risk of mastitis, an infection of the udder.

By cleaning your hands and the animal's teats at the beginning of the milking process each time, as well as dipping the teats afterward in an antimicrobial tincture, mastitis need not trouble you. Pathogens can lurk in soil, however, so I recommend purchasing a California mastitis test (CMT) kit so that you can detect early signs and preempt an infection before it becomes serious. You can purchase a testing kit through a dairy-supply outlet, and many extension offices have helpful information about testing for, preventing, and treating mastitis.

Test for mastitis at least once a week. The test kit comes with a paddle with four shallow cups; one cup for each teat. Start by stripping the first few squirts of milk from each teat; do not squirt this milk into the testing cups. Next, squirt a little milk from each teat into its corresponding cup on the testing paddle. Add a few drops of reagent to each cup and then gently swirl the contents of the cups by moving the paddle with a back-and-forth rocking motion of your wrist. If the animal has the beginning stages of mastitis, the milk/reagent solution will begin gelling. In advanced cases, it will instantly clump together. There are a number of treatments for mastitis specific to the type of infectious bacteria, so contact a veterinarian if an animal shows positive results on the CMT.

Milk: Raw or Pasteurized?

Many people worry about drinking unpasteurized milk. This is due in large part to decades of advertising. In reality, there has been a huge debate between the milk industry and the raw or "real" milk proponents for many years.

As people began moving from small farms to more urban areas, they still wanted milk, but milk was no longer in their backyards. So milk companies sprang up, and soon larger and larger dairy corporations developed, stepping in to fill the demand for milk. These large corporations often would often pool hundreds of cows in feedlot-type facilities, where it was nearly impossible to monitor individual animals for mastitis or other illnesses. Further, milk generated at many different facilities was transported to central processing plants, which means that milk from a variety of farms and cows (and cleanliness/pathogen levels) were combined into one huge lot for processing and then bottled for us to drink. Yikes!

Straight from the farm or pasteurized? There are arguments to both sides.

While it is true that in most, if not all, states, dairies must pass inspections to sell their products, the allowed level of bacteria is much higher than you would like to believe and certainly higher than what you'd find on your average one- or two-cow farm, especially if you are routinely test for mastitis. In order to keep the milk from spoiling quickly and people from becoming ill due to the bacteria in the multiple-dairy milk, companies began to practice and promote pasteurization, which heats the milk to 145–150 degrees Fahrenheit for half an hour.

If you research the pasteurization process, you'll find that heating the milk kills not only harmful bacteria but also healthy bacteria, enzymes, lactic acid, 80 percent of the available iodine, and other components, changing milk from one of the most healthy things that you can drink into a substance that triggers allergies, turns lactose (milk sugar) into beta-lactose, and renders insoluble most of the calcium content.

Few small farmers pasteurize, though home pasteurizers are available for purchase (they are not cheap). Instead, hobby farmers typically pay close attention to animal health and maintain cleanliness as a major component of their milking routines, from the preparation of the buckets, strainers, and bottles right through to straining and storing the milk or dairy products.

Milk from the store is not only pasteurized but also homogenized. Homogenization is a process of breaking up the fat globules found in cream to such a small size that they remain suspended evenly in the milk, rather than separating and floating to the surface.

You don't need special equipment to make small batches of butter with your fresh milk.

Since you haven't homogenized your milk after it comes from the animal, the cream will separate and rise to the top. Depending on the breed of your cow (Jerseys being among the richest in butterfat and Holsteins the least), the layer of cream can be 3, 4, even 6 inches deep on a gallon jar! There is a certain joy in seeing that solid creamy goodness on top of your own milk. You can stir the cream back into the milk, making it "whole" milk (or, as the British call it, "full cream" milk). Or, if you'd rather drink 2-percent milk, skim the cream and save it for use in coffee or for making butter or whipped cream. (Again, store in glass or stainless steel to extend storage life).

Butter

You can prepare butter as sweet-cream and sour-cream types. "Sweet" cream is not cream with sugar added. Rather, it is simply cream that has not yet soured, and butter made from it has a much milder flavor than sour-cream butter. (Store-bought butter is sweet-cream butter). However, it takes longer for the fat particles in sweet cream to coagulate than if you allow the cream to "sour" at room temperature for a day. One quart of cream will yield about one pound of butter and half a quart of buttermilk. Cream that has aged for several days in the fridge makes better butter than cream that just came from the critter (though it is still considered sweet if it hasn't begun to take on a tang), and will not taste as strong as cream aged at room temperature.

You have probably seen butter churns in antique stores. They are still available from farm suppliers, though they are rather expensive, especially if you won't be preparing a lot of butter regularly. For small amounts of your own butter for home use, the method I use is easy and quick and yields about ¼ pound of butter in less than half an hour.

Use a wide-mouth, pint-size canning jar, or, if you have the cream, use a quart jar for greater yield for the same effort (you can reuse clean canning lids for this application). Fill the jar about halfway so there is plenty of room for agitation. Add a dash of salt (up to a teaspoon depending on taste) and then begin shaking the jar to agitate the cream (other methods suggest adding salt after the washing process [see below]); in this case, you would not use a teaspoon full. You would salt to taste).

Agitating incorporates air into the cream and causes it to increase in volume so that the butter granules gather together. It is best done when the cream's temperature is between 54 to 64 degrees Fahrenheit. At first, the cream begins to take up the entire volume (classic whipped cream), but keep shaking and you'll find that, quite suddenly, butter granules begin to form and buttermilk separates out. When the granules form, continue shaking but not as vigorously, or you will whip the butter, making it more difficult to handle. When the butter granules have become gruel-like, they begin to clump together until you have a solid blob of butter swimming in buttermilk. Pour off the buttermilk (pigs, chickens, dogs, or cats will enjoy this treat) and then wash the butter with cold water, working the lump with a wooden spoon to release hidden pockets of buttermilk that, if not removed, will cause the butter to sour more quickly. Keep washing and working with the spoon several more times until the liquid you pour off is fairly clear. If you did not salt the cream in the beginning, now's the time to work salt into the washed butter—about ⅛–¼ teaspoon to taste.

Hand-form the butter into blocks, pack it into a plastic storage container with a lid, or, for a more traditional approach, press it into a mold, chill, and then remove from the mold.

Using a kitchen mixer also works to make butter. Anyone who has beaten whipping cream for too long knows that it turns grainy. Go a little further, and you've got butter. One caveat, though: if you're using a stand mixer, don't leave it unattended at the butter stage because once butter begins to stick to itself, it will clump on the beater and begin to slosh the buttermilk all over your kitchen. This is the voice of experience speaking!

Add fresh fruit to plain homemade yogurt for a delicious, healthful snack.

Yogurt

You can make yogurt by adding *Lactobacillus bulgaricus* and *Streptococcus thermophilus* to fresh milk (some yogurts also contain *L. acidophilus*). These bacteria produce lactic acid during fermentation, which lowers the pH and causes the milk protein to thicken. This partial digestion of the milk makes yogurt easily digestible, and the resulting cultures help maintain healthy intestinal flora.

One of the factors that is crucial for successful yogurt-making is the use of sterile techniques, including properly cleaning all tools and containers and keeping unwanted bacteria away. You must also maintain the incubation temperature between 100 and 122 degrees Fahrenheit; temperatures within this range inhibit the growth of pathogenic bacteria yet are low enough to retain the yogurt's nutritive value. This temperature range is also ideal for the good bacteria that do the fermenting and for protection of the starter from contamination. Starter is just plain yogurt, unsweetened and unflavored, either store-bought or from a previous batch of homemade. The degree of tartness of your finished yogurt depends on how long you ferment it. At 100 degrees, mild yogurt takes about six to eight hours, or about three to four hours at 122 degrees.

Yogurt in Baking

Baked goods rise well when you use yogurt due to its acidity. When you use yogurt as part or all of the liquid in cakes, waffles, pancakes, and muffins, cut down on the amount of baking powder accordingly.

To prepare yogurt, warm 4 cups of milk in a pot over medium heat until it reaches 122 degrees (for richer yogurt, substitute ¼ cup of heavy cream for ¼ cup of milk).

Ladle about ½ cup of the heated milk into a bowl or your measuring cup. Add ½ cup of starter and mix thoroughly. Next, return the starter/milk mix to the pot and stir until you've distributed the starter well. Pour the mix into sterilized canning jars (you can add chopped, canned fruit and a little syrup—or other flavors—to the bottom of the jar before adding the yogurt) and cover with new canning lids. Although making yogurt is a most-of-the-day job (it is a great project for rainy or snowy days), once it is in the jars, it isn't too labor-intensive. Keep the mix at a temperature between 100 and 122 degrees Fahrenheit until it has fermented. The way I do this is by setting the yogurt jars in my canning kettle after the water in the kettle has reached about 122 degrees. In the winter, I place the pot on the wood stove and in the summer, on a burner on the cook stove at a low setting that will hold the water at 122 degrees. High temperatures kill yogurt cultures.

I take the kettle off the heat source until the temperature gets down to about 100 degrees and then reheat it to 122 degrees, repeating this step until the mix turns to yogurt. It usually takes only a short time to bring the kettle up to 122 degrees, and it takes a couple of hours to get it down to about 100 degrees if I put the lid on to help trap the interior heat. For a typical batch, I might have to reheat the kettle three times during the day. Just set a timer so you remember to check it. Another option for yogurt-making is a home yogurt incubation unit, which comes complete with containers, lids, and a heating base that will maintain your temperature for you. After fermentation, clean the jars and store them in the fridge.

If your only experience with yogurt is the kind you buy in the store, the first thing you'll notice the whey, a pale liquid that forms on top of your homemade yogurt, and the layer of cream. Whey is just the water fraction of the milk. You can purchase a strainer if you want to get rid of all the whey, but before I eat the yogurt, I usually just pour the whey off and give it to my dogs, who love it. If you mix the yogurt and don't eat it all in one sitting, you'll see that more whey has separated. Just pour it off again.

I generally make plain yogurt and add fruit, nuts, or flavors as I use it. Plain is great for baking (substitute yogurt for up to half of the liquid in your favorite recipes or in dips and sauces that call for sour cream. The recipe yields two 1-pint jars of plain yogurt.

You can also make yogurt from store-bought milk if you don't yet have your own cow or goat. In this case, the layer of cream will not rise, but all else should act about the same. Yogurt made from raw (unpasteurized) milk will, of course, contain more nutritional value.

Cheese

You'll find different types of cheese and cheese-making techniques throughout the world. Cheese can be made not only from cow's or goat's milk but also from the milk of sheep, horses, buffaloes, or yaks, and while you may not have all of these animals on your farm, you can make a wide variety of cheeses if you have the extra milk. You can even make cheese from store-bought milk.

All cheeses fall into one of three categories: hard, soft, or cottage cheese. Hard cheeses (such as cheddar) require aging and are more complicated to make, but they can be stored for long periods of time. You make soft cheeses the same way as hard cheeses, but soft cheeses are aged for a shorter time. Cottage cheese is a soft cheese that you do not allow to age. You can easily make soft and cottage cheeses at home without too much specialized equipment, and they have a refrigerator life of five to ten days. Hard cheeses take a bit more equipment and time, but many people make them at home with success.

Homemade soft cheese, flavored with herbs and spices.

The basic ingredients of cheese are:

- **Milk:** You can make cheese from whole milk, 2 percent milk, or skim milk, though the cheese's richness correlates to the amount of milk fat.
- **Rennet:** An enzyme—rennin—found in rennet converts the protein, or casein, in milk from a soluble form to an insoluble form, causing the milk to clump. The rennet commonly available in supermarkets can work for the softer cheeses, but for hard cheeses, you'll get better results using rennet specifically for cheese making. These animal- or vegetable-based products can be ordered online from a cheese-maker's supply house (see resource list for suggestions).
- **Starter:** You must add bacteria to acidify the milk so that the rennet will work as well as to aid in the curing. Cultured buttermilk or yogurt can serve as a starter for some types of cheeses, or you can purchase pure cultures from a cheese-maker's supply house. Cheese-making recipes will indicate what type of starter you need for particular cheeses.

Meat

Butchering involves the actual slaughtering of animals and cutting up the carcasses for meat. Processing is taking raw meat and making some other type of product through the addition of other ingredients (such as beans, tomatoes, onions, peppers, and spices for chili) or through specialized handling, such as smoking. With the exception of poultry (which you can generally butcher on the farm), I'm not going to go into the actual processes involved in butchering and processing. Most communities have one or more meat-processing businesses that will help you from start to finish or will do the parts of the butchering that you do not want to do yourself. Big animals are more difficult to butcher on the farm, but as multitudes of deer hunters will tell you, it is definitely doable, if time-consuming. Actually, until the 1950s or later, most farmers butchered their own beef and hogs, and certainly our agrarian grandparents or great-grandparents were proficient at it.

Depending on the type of animal and how old it is, you can either butcher it immediately or age the meat for up to two weeks by hanging it in a cooler (or butcher during late fall or late winter/early spring when temperatures fall below forty degrees but hover just above freezing). Aging helps cure meat, improving flavor, texture, and tenderness.

If you want to learn the techniques of how to butcher your own large animals, get a copy of Dr. John Mettler's book, *Basic Butchering of Livestock and Game* (see Resources) or check with your extension agent for a booklet on butchering.

What I do want to go over are some things you'll need to know about butchering and processing if you plan to raise animals for meat. For instance, how much meat will a 1,000-pound steer actually yield? How do you respond when the butcher asks how you want your meat cut up? Roasts? Steaks? Hamburgers? What do you say to a first-time customer who is considering purchasing a lamb but wants to know how much meat he'll receive and how much freezer space it will take up?

When a butcher uses the term "yield," he or she is generally referring to the *hanging weight*. The term comes from the fact that a butcher hangs a carcass on a rail attached to the ceiling of the cutting room to make his work easier. Typical hanging yields are:

Find a good local butcher if processing your own meat is not something you want to do.

- Lamb—50 to 55 percent of live weight
- Beef—60 to 65 percent of live weight
- Pork—70 to 75 percent of live weight

The meat yield depends on how closely you trim the fat and on how many "bone-in" cuts you prepare. Typical meat yields are:

- Lamb—40 to 45 percent of live weight
- Beef—45 to 50 percent of live weight
- Pork—55 to 60 percent of live weight

For example, a 250-pound hog will hang on the rail at about 185 pounds, and when you pick up the packages of meat, you will receive about 140 pounds of meat. A 1,000-pound grass-finished steer will hang around 650 pounds and yield 500 pounds of meat.

On average, cut and wrapped meat requires one cubic foot of freezer space for each thirty-five pounds of meat. The meat from a whole steer will almost fill a fifteen-cubic-foot chest freezer. The pig will require four cubic feet, which is about the size of the freezer space in a large combination refrigerator/freezer, but it is too much for a smaller combo unit. The lamb will fit into the freezer compartment of a smaller refrigerator/freezer.

When you drop off an animal at the butcher's shop, the butcher will want to know your "cutting orders." The cutting orders are simply the directions that tell him or her how you want the animal cut and wrapped. The first time you figure this out, it can be a bit mind-boggling. Thankfully, most butchers will have a list of possibilities/suggested cuts. A good way to figure out what will fit your needs is to look at the way you currently use meat and then order accordingly. If you use more hamburger than roasts, consider turning some of the less prime roast cuts to burger. If you normally cook ground meat in one-pound increments, order it that way. Each processor will have his or her own suggested packaging sizes but are usually willing to customize for you.

As you use up your meat, assess how well your cutting order is working out for your family. Do you run out of roasts too soon or wish you'd saved a few more steaks? How is your hamburger supply holding out? This information will help you the next time around. The good news is that all of the meat is edible, so it's a win-win situation no matter what form you put it in. If you end up with some unfamiliar cuts, you will either know what not to order next time, or you will have discovered a new favorite cut!

The first thing to tell your butcher is how you want the animal trimmed (cutting away external fat). We go for well-trimmed meat, and our ground beef is 95 percent lean. At the same time, some processed meat products do benefit from the addition of some extra fat from the trimmings. One time we had some all-beef wieners prepared with no additional fat added, and they were too dry. Luckily, Karen, the woman who took cutting orders at the time, warned us that all-beef wieners might be too dry, so we only had a few pounds made up to try them.

The butcher also needs to know how many steaks or chops you want in each package and how thick to cut them. We have found—particularly if you plan to direct-market your meat—that packages with two steaks or chops are a good size. We have also found that that ¾-inch-thick cuts seem to be ideal for most steaks and chops. Most of the meat from the round (the butt) and the chuck (the shoulder) is best ground or in roasts, though you can do some amazing things in the culinary realm with a humble chuck or round steak.

The first time we took a steer in to have it butchered, Karen asked if we wanted the short ribs. I stared rather dumbly and said, "Well, I don't know. Why?" She went on to explain that most people didn't take the short ribs because they're really tough and stringy and there isn't much meat on them. I didn't take them that first time, but the next time we took in a steer, I decided to keep them. By this point, I was thinking they'd at least be dog food if nothing else. I also told her we wanted all of the organ meat—even the tongue—as well as packages of soup bones and boxes of bones for our dogs. Our policy became "get everything back, because something will eat it."

A butcher will process your meat according to what cuts and products you want.

Short ribs are, true to their reputation, stringy and tough when prepared by most conventional cooking methods. But we discovered that they're actually pretty darn good if cooked in a pressure cooker. We put about an inch of water in the bottom of the pressure cooker, load in the short ribs, pour

Various types of smoked pork, including chorizo sausage, ham, and salami.

some barbecue sauce over the top, and pressure-cook them for about forty-five minutes. After this treatment, they fall off the bone and taste great even though they are messy to eat!

Lamb, bison, deer, and elk all have similar cutting orders to beef, but pork is different. The butcher will want to know if you want hams or fresh roasts. Hams are cured (smoked) roasts. Bacon is cured "side pork." You can make the front shoulder of a hog into a picnic ham or slice it for pork-shoulder bacon. Pork-shoulder bacon is nice lean meat, much like Canadian bacon. There is lots of "trim" on a hog (and it isn't all fat, so it can be used for sausage jerky).

Butchering Poultry

Butchering poultry isn't fun (and for the more squeamish or for those whose birds have names and personalities, you might just want to leave it to someone else), but this is one job that farmers usually do on site if they raise chickens for meat. If you're planning to butcher many birds, you need to develop an assembly-line approach to butchering (or find a poultry packer), but when starting out with a small number of birds for home use or to begin assessing your production and marketing strategies for a commercial enterprise, the following directions work. Beware: Graphic content. I apologize ahead of time, but for such a process, a certain number of details are necessary.

A butcher can make specialty meats, such as fresh sausages.

1. It is best to remove feed from the birds you plan to butcher 12 to 24 hours beforehand. This helps empty the digestive tract so you have less mess to deal with. They should still have access to water and should be kept to their normal routine in all other ways so that they are not stressed.

2. There are many ways to actually dispatch the bird. Though killing any animal isn't a pleasant task, each of these methods, when done correctly, is quick and humane: wringing the neck, chopping off the head, or slitting the throat (see sidebar on page 261).

3. After you kill the bird, allow it to "bleed out" for a couple of minutes. By draining the blood, the meat remains cleaner and less likely to spoil. One way to drain the blood is to tie the feet together and loop the twine over a fencepost.

4. Scalding loosens the feathers, making plucking easier. You can do this outside to minimize the mess in your kitchen. In a bucket or kettle big enough to submerge the bird without overflowing the rim (if you're inside), bring clean water up to 130 degrees Fahrenheit for a

small bird or 140 degrees for a large bird. If the water is not hot enough, the feathers will not be easy to pull. Submerge the bird in the scald bucket for about 40 seconds by holding it by the feet and dunking it headfirst. If your scald water gets a little too cool, you'll need to leave the bird in longer; if it gets too hot, the bird will be overscalded, which results in "cooked" meat. Swishing the bird around a little in the bucket helps to move the heat through all the feathers down to the skin, causing it to release its grip on the feathers. When doing multiple birds, try to keep the

scald water fairly clean, as it can be a source of contamination. If the birds are dirty, hose them off prior to scalding.

5. Pick feathers immediately after scalding, as soon as the bird is cool enough to handle. When plucking by hand, it's easiest if you hang the bird back up so that you're working at a comfortable height. (Sorry to be the bearer of bad news, but this part really stinks. Literally. Wet chicken feathers smell worse than wet dog fur, so prepare yourself).

6. Use rubber gloves to give you a better grip (surgical gloves that fit tightly work best) and pull the feathers down and away from the body. If the scalding temperature was right, most of the feathers will come off fairly easily. If you plan on butchering a lot of birds, a rubber-fingered "plucker" makes the job easier. After picking most of the feathers, rinse the bird with cool water to reveal any remaining feathers and the pinfeathers. Don't rinse in a bucket, but with a garden hose or under a faucet in order to cleanly flush the carcass.

7. With a short, dull knife, scrape off the pinfeathers (feathers just starting to grow that don't always come out while plucking). You can also use tweezers for this purpose. Singe any hairs with a propane torch or a twist of newspaper lit with a match (older birds have more hairs than young ones do). A method that is commonly used with waterfowl (they have lots of down and pinfeathers) is to dip the bird in hot wax (150 degrees Fahrenheit), then dip it in cold water to set the wax, and finally peel the wax away. The bird may require a second dip to remove any remaining hairs.

From this point forward, you may find it easier to work on a cutting table than with the bird hanging. Try both approaches. Again, in this age of Internet, you can watch videos on how to process a chicken from start to finish once it has been plucked. Sharpen all knives to prevent having to use excessive force. Here is a basic rundown on the steps:

1. Remove the head by cutting just below the first vertebra in the neck and then twisting it off.

2. To remove the neck, first insert a thin, sharp knife (a filleting knife works well) into the skin above the neck, just at the shoulders, and cut forward to open the skin, Pull the skin away from the neck. Next, pull the crop, the trachea (windpipe), and the gullet away from the

neck skin, and cut off where they enter the body. Be very careful not to cut into the crop, as it can leak partially digested food onto the meat (if this happens, wash it off immediately). Finally, cut the neck off where it enters the shoulders by cutting the muscle around the bone and then twisting it off. Wash the neck and then set it aside in a bucket of ice water.

3. About 1 inch in front of the oil "nipple" on the tail, make a clean cut all the way down to the vertebra in the tail. Cut out the oil gland from front to back and scoop out the gland.

4. Remove the shank (the lower part of the leg and foot) by cutting through the hock joint from the inside surface.

5. The next step is evisceration (removal of the animal's internal organs). There are two methods: one that is appropriate for small birds that you won't truss for roasting, and another that is appropriate for large roasting birds, such as turkeys. This process takes practice in order to become adept, so take your time when learning. You might watch a few videos on the subject or help a few neighbors do their own butchering. If you accidentally

Evisceration Methods

For smaller birds: Starting just below the point of the breastbone, insert the thin knife just enough to penetrate the skin and muscle. Cut down toward the vent (anus) and then cut all the way around it. When cutting around the vent, keep the knife pressed as close as possible to the back and tail. Pull the vent and a small section of the large intestine out of the way.

For larger birds: Make a half-moon cut around the vent, pressing closely to the tail and backbone. Insert your index finger into the cut and up and over the intestines. Pulling the intestines down and out of the way, continue to cut around the vent until you have completed the circle. Pull the vent and a short section of the intestines out of the way. Next, pull the skin back toward the breastbone and then make a side-to-side cut about 3 inches wide and 2 inches below the breastbone. Finally, pull the bar of skin that remains backward and over the piece of vent and intestines.

Methods for Killing a Chicken

To wring a bird's neck, pick it up by its head and swing it completely around in a 360-degree circle. If you use this method, hang up the bird after it is dead and cut its throat, as described below.

Chopping off the head with an ax or hatchet on a block is an old, though somewhat messy, method. For a modern variation that is a bit easier, drive two large nails into one end of a length of 2x4. Place them about ½ inch apart in a "V." Next, place the chicken's head in the V to hold it in place, stretch the neck out, and give a sharp, strong blow with the hatchet about 2 inches from the head. Let go of the chicken, which will flop about for a few seconds. Be comforted that this method is quick, and the bird is feeling no pain.

Another common method is to hang the bird upside down with its feet tied or in a killing cone. Using a sharp knife, slit the bird's throat by making a cut directly behind the lower jaw, but try to avoid cutting through the esophagus and windpipe. When you cut the bird's throat, you cut the jugular vein, so a bucket underneath the bird will catch the blood and make cleanup easier. This method calls for a bit more skill, however, and is not as quick or painless for the bird.

cut through the intestines, releasing their contents onto the meat, don't panic. Just wash the meat immediately and thoroughly.

6. For smaller birds (e.g., chickens and ducks), insert your hand into the abdominal opening and begin gently working your way up and around, loosening organs from the body cavity as you go. When all attachments are broken, scoop the insides out. Sometimes even if you haven't cut into the intestines, this pressure will cause the contents to leak out. If this happens, wash it off/out of the carcass immediately.

7. If you want to save the giblets, remove the gizzard, liver, and heart from the rest of the entrails that you've just pulled out. Cut the gizzard away from the stomach and intestines. Peel away the fat, split it open, rinse it well, and then add it to the ice water. Trim off the heart sac and heavy vessels from around heart, rinse, and add the heart to the ice water. Pinch the gallbladder off of the liver, rinse the liver, and add it to the ice water (discard gallbladder). Retrieve the gizzard from the ice water and peel away the lining by inserting a fingernail under the lining at the cut edge and pulling away. Return the gizzard to the ice water.

8. Use your hands to check along the backbone for any remaining organs (the lungs and sexual organs are often still in the bird). Remove any that you find. Rinse the bird well under running water, both inside the body cavity and on the outside of the carcass. When the carcass is clean, put it in your ice-water bucket to chill. Adding a few tablespoons of salt to the water will help pull any blood still remaining in the tissue out of the meat and into the water.

9. Chilling time varies, depending on the size of the bird and the temperature of the water, but it may take several hours. The goal is to get the carcass temperature down to about 40 degrees Fahrenheit as quickly as possible. Once the carcass is adequately chilled, rinse it once more and then hang it up to drain for about ten minutes. While the carcass is draining, wrap the heart, liver, and gizzard in plastic wrap. Insert the neck and the wrapped organs into the carcass.

10. Finally, bag the bird, label, and refrigerate or freeze as desired.

11

Agripreneurship

First, the disclaimer for this chapter: More often than not, new and aspiring hobby farmers harbor a hope that they can make some money from their operations. Maybe they just hope to sell a few foals from their brood mares to help pay for the feed and vet bills, or maybe they dream of progressing from a hobby farm to a fully functioning value-added farming operation that will support their families in some level of comfort and security without needing to hold outside jobs.

You probably remember the Kevin Costner film *Field of Dreams*. In it, Costner plays a baseball-loving farmer who creates a baseball diamond in his cornfield. Needless to say, everyone thinks he is crazy—and perhaps he is—but his "if you build it, they will come" attitude somehow carries the story through to a fulfilling ending. It's a nice story, but real life can hand us quite another scenario.

As people who have gone into business for themselves, be it farming or some other type of small business, will tell you, it's easy to start your entrepreneurial journey with a *Field of Dreams* mindset: "build the business, and customers will come." True, start-ups can be exciting, carried along by dreams and initial energy, but attracting a customer base takes research, planning, and hard work. While adhering to any regulations that apply, you need to create a marketable product. That marketable product must be followed by promotion, promotion, promotion—and more promotion. Even then, it can be a hard row to hoe.

If you think that you can raise a few foals and make a profit suitable to keep the family comfortable on a single job, or if you hope that a 1-acre

Handmade herbal products are popular for their natural and artisan qualities.

market garden or a community supported agriculture (CSA) venture will, in and of itself, result in a lifestyle of comfort and security while providing the world with wonderful veggies, think again. I hate to rain on your parade, but the reality is that very few start-ups yield the equivalent of a real-world salary, at least at first.

That said, people do it all the time. Those who are willing to do the work of developing value-added or alternative-enterprise approaches have the potential to make small farms into successful and sometimes very profitable businesses. There are plenty of examples out there of people doing innovative things in and around agriculture—and doing them quite well. They are making good incomes doing something they love.

The common thread in the success stories all highlight people who have figured out how to deliver what consumers want. For example, consumers are:

- looking for local, fresh, organically or naturally grown products and are often willing to pay extra for them;
- more interested in how and where their food is raised and produced and how that production affects their health, the environment, and society;
- charmed by well-made farm-based artisan products such as lavender soaps, goat's milk cheeses, jams and preserves, hand-knit woolen sweaters, and more; and
- looking for opportunities that small farms can provide, such as educational activities, outdoor recreation, and agritourism.

The Market for Farmers

Traditional farmers raise commodity products, such as corn, soybeans, beef, and pork, and sell them through the commodity distribution system. For example, a farmer will sell livestock at a local sale barn or sell grain to the local elevator. If a farmer is producing on a large enough scale, he may get into the futures market to try to increase profits, but this only works for the big guys. In the commodities game, external forces set prices, and you can't do much to differentiate yourself or your product. Commodity producers continually struggle to sell enough product to make ends meet.

Selling value-added products, or running an alternative enterprise, offers the opportunity to increase income by gaining a bigger share of the consumer's dollar. We define value-added products as those that:

- are changed in physical state or form, such as turning berries into jam or creating sweaters and hats from wool;

- are produced in a manner that enhances value, such as products that meet organic or humane standards; or
- result in the enhancement of the value of a commodity or product (for example, creating branded meat or dairy products).

Alternative enterprises include direct marketing to consumers, raising crops that don't relate to traditional commodity crops (e.g., Christmas trees, medicinal herbs, or colored eggs), producing renewable energy from wind or sun, and providing activities that bring people to your farm (e.g., educational events, hayrides and corn mazes, hunting or fishing, U-pick berries, or bed and breakfast accommodations).

Farmers engaging in value-added and alternative enterprises often build relationships with customers—customers who aren't faceless masses who live a thousand miles away, but individuals who count on you to be a safe and trusted source of high-quality, unique products. Going into an alternative marketing system will usually require you to follow additional laws and will always require extra planning, record keeping, and time.

Getting the word out to potential consumers is a major concern for farmers pursuing alternative markets. Educating your customers about your product, why your product is special, and why they should buy from you becomes an important part of your job. The payoff can make the extra effort worthwhile, but revenue rarely pours in overnight. Like so many things on the farm, your business will not erupt full-sized and fruitful. It must be grown. Tended. Developed. Encouraged along the way.

Niche marketing requires you to differentiate your product from the tons of products that consumers look at when they wander the aisles of the local superstore. No matter what product you're selling from your farm, if you're going to direct market it, you will need to think about how it will best fit in a niche. Some niches include:

- Natural: The US Department of Agriculture (USDA) allows the word "natural" to appear on the labels of minimally processed food products that contain no artificial colors, flavors, preservatives, or other additives.
- Organic: The organic market is the fastest growing segment of the US food economy. To qualify as organic, you must grow produce and raise animals according to organic guidelines, using only approved substances and methods. For example, the organic market strictly prohibits the use of chemical pesticides, but it allows diatomaceous earth for insect control and as a wormer for certain livestock.
- Green/humane: Products produced and marketed as "green" or "humane" appeal to consumers who want to know that they are supporting environmentally friendly farming practices and "happy and healthy" animal husbandry.

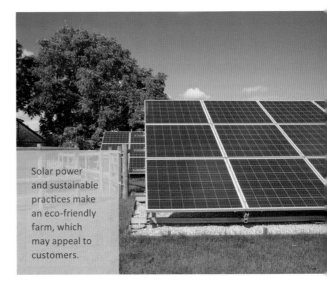

Solar power and sustainable practices make an eco-friendly farm, which may appeal to customers.

Legal Issues

When you go into business, you will fall under a bevy of rules and regulations promulgated by federal, state, and local government entities designed to protect the public. For each business idea you consider, you must explore the legal concerns that may arise: Do you need a license to establish the business? With what kind of regulations will you need to comply? Wandering through a maze of red tape and bureaucracies can be, at the very least, intimidating and frustrating, and, at worst, a nightmare. But don't give up! You can do it, and the payoff is worth the trouble.

Understanding your rights and responsibilities is crucial to successful alternative marketing. To find out what laws you might collide with, a good place to start is your local library. Librarians can point you in the direction of the reference shelves housing books of statutes and ordinances. Librarians are also generally happy to help you figure out how to find the information you need if it's your first time reading these onerous documents. Reference materials must usually stay in the library, so bring change for the copy machine. Also, get on your computer because all federal laws and some state laws are available via the Internet, and more and more local governments are posting their laws online as well. Find links to federal and state laws on the Internet at a site run by the US government: www.usa.gov. Click on "Services and Information" and navigate to the Reference Center to find a link to federal and state regulations.

If you plan to employ nonfamily workers, you'll have additional requirements, such as acquiring an employer tax identification number from the Internal Revenue Service, filing payroll taxes, and registering for both unemployment insurance and worker's compensation insurance. You must file payroll taxes on a monthly or quarterly basis, depending on the size of your payroll.

The Law in Practice

Let's look at a few scenarios to give you an idea of how laws work in real-life situations. Your accountant or tax preparer can also be a great resource for these types of issues.

Scenario 1: You want to start an on-farm bed and breakfast. In this case, there are probably no federal laws to worry about. You aren't carrying on interstate commerce because your product can't possibly cross state lines, so federal business laws don't apply. And you probably aren't engaging in any practices that could violate other federal laws; for example, you aren't processing foods or drugs, and you aren't spewing out air pollutants or water pollutants.

Business Resource

A great resource of online materials for agricultural business development is the Agricultural Marketing Resource Center, found online at www.agmrc.org. They have sample business plans and a spreadsheet called the "feasibility template" that can help you develop your plan.

In this scenario, the only laws you need to worry about are those adopted by local governments or your state. How do you find out about them? Start with the local municipal code for your town or county. Look in the table of contents or the index for headings such as "business, licensing, lodging, and guest houses." Read each relevant reference carefully, and document all of the statutes that apply, even if only remotely, to your proposed operation. If there are no local laws that apply to what you want to do, then check the state's statutes.

A bed and breakfast will need various licenses and permits for which you'll need to fill out applications and remit fees. Application forms may require sign-offs from certain government departments, such as the local fire department or health department. You'll need to have your septic system inspected and approved to handle increased occupancy, and you'll need to insure your business, either through a rider on your homeowner's policy or a separate policy. You'll be cooking and serving food to your guests, which brings into play laws regarding food preparation and handling. So make sure that you find out what is required by all state and local government entities that have any jurisdiction over your site and type of business, including fire districts, water and sewer districts, and

town or county governments. There are many organizations, such as the Professional Association of Innkeepers International (www.innkeeping.com) that help members and aspiring entrepreneurs explore how to pursue their B&B dreams. In addition, you can contact each government entity—town, county, state—and talk to staff members about what you have to do to start your bed and breakfast. As they tell you what hoops you have to jump through, be sure to make note of the specific laws to which they are referring so that you don't end up doing unnecessary work.

Scenario 2: You want to direct market beef (or lamb, or pork, or goat), which falls under the United States Code (a compendium of all federal laws). The part of the US Code that deals with meat is Title 21—Food and Drugs, under which Chapter 12, Meat Inspection, details the requirements for a federal inspector to examine all carcasses intended for use as meat or meat products. The intent of such inspections is to prevent the use in commerce of meat and meat food products that are "unwholesome, adulterated, misbranded." Interestingly enough, if you direct market bison, deer, or elk, you'll notice that they don't fall under this law because they weren't included when the law was written; however, other federal laws apply to marketing meat from these animals.

You can sell meat in one of two ways: "on the hoof" or packaged. Selling on the hoof means that you are selling the live animal to the customer, and he or she is then responsible for having it butchered to his or her specifications. Selling meat on the hoof requires less red tape than selling packaged meat because there is a specific exemption (Title 21, United States Code, Section 623) that allows the owner of an animal to have it butchered for personal use without having it inspected. This means that the buyer can have the animal butchered at a "custom slaughter" facility as opposed to a federally inspected plant. A federal or state inspector must occasionally check custom plants for cleanliness, but, unlike federally inspected plants, there are no inspectors checking the carcasses of each butchered animal. Generally, many smaller packing plants fall into the custom category, which is a blessing; with their small scale, they can't afford to pay for a regular federal inspector to be on site.

You may split an on-the-hoof sale among multiple customers—for example, two families could each be buying one half of a hog from you—but you must charge them based on the live weight of the animal, and they must pick the meat up from the butcher themselves. When you meet these criteria (selling based on live weight, butchering per customer specifications, and requiring customer to pick up meat), the exemption from carcass inspection kicks in.

If you decide to sell packaged meat (cut and frozen or processed into meat products, such as jerky and sausage), you must slaughter each animal at a plant that has a federal inspector available to view the carcass. You must label packages, detailing what the product is and who distributes it. If the package contains a product that has any added ingredients whatsoever (even just a pinch of salt), the label must list these ingredients. The USDA must approve the label, and the USDA symbol must appear on the label. You must be able to support all claims made on the label. Federal law also requires that all meat packages carry a "safe handling instructions" label.

After you have cut, processed, frozen, and labeled your meat, you must follow laws that apply to how it is stored and transported. Finally, check to see if any local laws apply to what you're trying to do. For example, if you want to open a retail meat store on your farm, you may have to go through the planning, zoning, or building department for approval.

Scenario 3: You want to sell chickens or eggs. Chickens and eggs, like the meat in the second scenario, also fall under Title 21, but there are generous exemptions for small-scale producers who wish to direct market. The exemption for poultry allows a complete exemption from inspection for any producer who raises and slaughters fewer than 1,000 birds per year. Producers who slaughter between 1,000 and 20,000 birds per year are exempt if they market the birds within their local jurisdictions. This exemption not only applies to poultry you have raised and slaughtered on your own farm but also to birds that you sell to commercial outlets—such as restaurants and hotels—as well as those birds that you sell directly to consumers.

In the case of eggs, the exemption from inspection extends to producers with fewer than 3,000 laying hens, but there are limits on how many eggs each customer can buy; these vary from state to state. Small producers are still required to label their egg cartons with their name, address, the quantity of eggs, and the size of the eggs. It's also a good idea to date the carton.

Although most states generally accept the federal standards, some state and local government entities have adopted more stringent requirements than the federal requirements for meat, chicken,

Sellers of eggs straight from the farm must follow certain guidelines regarding labeling.

and eggs; for example, some states require small egg producers to have their eggs candled and graded by a licensed candler. It is your responsibility to know which laws apply to you. An old adage in the legal world says that "ignorance of the law is no excuse of the law." So, play it safe and check your state and local laws—such as county health department laws—to be sure that you are in compliance.

Be aware of the potential dangers that exist on your farm.

Liability

Unfortunately, we live in a time when there are a lot of lawyers, and lawyers need lawsuits, so liability is a serious issue. Liability generally arises when a person fails to meet his or her responsibilities. The best way to avoid personal liability is to pay attention to details. Make safety a top priority on your operation, studying each area of your farm and everything you do for possible hazards. Follow all laws; they may seem like a nuisance, but they offer you a degree of protection. If you have employees, develop written personnel and safety policies.

Safety policies can also extend to customers. Place signs where your customers will see them. Areas that are off limits to customers should be clearly marked; for example, "Do not enter the barn" or "Employees only beyond this point." Also, don't be afraid to tell customers or their children to stop doing something that may be dangerous: "Please do not pet the bull" or "Please stop chasing the chickens." If a customer is annoyed and leaves in a huff, count your blessings—you don't need people on your farm who aren't safety conscious and who don't respect your rules.

Depending on what you offer on site, you may want to have customers sign a waiver in order to visit. Even if you are very safety conscious, accidents still occur. And even though an accident may not be your fault, a customer may still sue you. Your property insurance company should be able to offer you extended personal liability coverage for both employees and customers, but these policies generally don't cover on-farm marketing operations, so check with your insurance agent about liability and loss insurance specifically designed for direct-market farmers.

Business Plans

Business planning is a process that will help you define your market, how you will best sell to that market, and how you will manage your operation. Drafting a business plan is a valuable process for anyone in business because it first helps a person think through his or her approach to the enterprise and then helps track, monitor, and evaluate the business's progress. A business plan is also absolutely critical if you intend to seek outside financing to get your business off the ground or to keep it growing because lending institutions or investors will want to know exactly what they're being asked to invest in and will want to see proof (at least on paper) that it is a viable enterprise. Your business plan may help you deal with things like the county zoning board or suppliers. A business plan typically includes four standard sections: a mission statement, a business description, a marketing plan, and a financial plan.

The mission statement should detail in a few sentences what is special about you and your business and what you are seeking to accomplish. Consider a few examples:

- The mission of Smith Farms is to provide wholesome, natural foods that are produced in an environmentally responsible manner for consumers in the Greater Cleveland area. We plan to market our produce through area farmers' markets and at a farmstead stand that will be open from May through September.

- Our mission at Whitehorse Farm Bed and Breakfast is to provide an experience for consumers who desire a chance to get back to the country. We provide great personal service to visitors, who have the opportunity to stay in one of several historic cabins or in the 1920 farmhouse. We've meticulously renovated all of these buildings, which are located on our working farm. During their visit, our guests have the chance to pet a cow, ride a horse, or shear a sheep. They can also fish in our stocked pond, hike on a series of trails along a river and through a native grassland that teems with wildlife and wildflowers, and feast on farm-fresh food of the highest quality.

- Kelly's Farm has long been known for its abundant high-quality fresh produce. Today, our mission is not only to supply these exceptional fruits and vegetables to our clientele but also to offer customers a variety of prepared foods, including Kelly's Red-Hot Salsa and Kelly's Cool Catsup, that will excite their taste buds. All of our products are market driven by the demands of discerning consumers who want the best, freshest ingredients in a locally grown and prepared product.

After you have written your mission statement, go into a more detailed description of the business. The business description should be several pages long and should provide as much detail as possible. It should tell the reader who the owners are and describe the owners' experience and/or education relevant to the endeavor. It should describe the type of business structure (sole proprietorship, partnership, corporation) and the history of the business if you have been operating for a while. It should also provide all other pertinent information about the business. Let's look at a hypothetical business description for Whitehorse Farm Bed and Breakfast.

In 1898, William Garry homesteaded Whitehorse Farm. Today, Gill and Jenny Garry and their family continue the tradition, operating Whitehorse Farm as an S corporation.

The farm is 800 acres and is located in Wayne County, about an hour from St. Louis. We raise and train American quarter horses that are well known in horse circles for riders who participate in cutting and reining events. We maintain a purebred Angus beef herd and a commercial cow/calf herd of mixed-breed animals. We also maintain a small sheep flock for meat and for fleece, which we sell to hand-spinners and weavers.

The bed and breakfast (B&B) will be operated by Jenny and by Tom and Karen Smith (the Garrys' daughter and son-in-law). Tom has a degree in business administration from the University of Nebraska, and Karen has worked for seven years in hotel management for a Holiday Inn in St. Louis. Jenny, whose cooking skills regularly win not only praise from family and friends but also awards at the Wayne County Fair, will provide meals to visitors.

Visitors to the B&B will choose between three cabins and the historic three-bedroom farmhouse, which also has a large room suitable for group events such as weddings, meetings, and parties. Cabins have small refrigerators and microwaves; the farmhouse has a full kitchen. We will supply all visitors with a daily breakfast of sweet rolls, buns, and fresh fruit, and we will offer catering for special events, such as corporate retreats.

Our objectives are to:

- *incorporate this complementary operation into our overall business;*
- *have at least 30 percent occupancy our first year in operation, and 50 percent occupancy our second year; and*
- *increase exposure and market using Internet technology and direct advertising in the St. Louis area.*

The next section, the marketing plan, expands on the information in the business description. Before you complete it, you will need to do market research to identify customer needs and wants and to determine whether the product or service you are considering meets those needs and wants. You'll also identify specific target markets (for example, senior citizens in Milwaukee or Asian Americans) and determine the best way to advertise your business to each customer group. Your marketing plan should also delve into who your competition is and how you are set apart from them, and it should provide information on how you will reach your target markets (for instance, through print advertising, social media, or cross-promotion with another business).

Market research looks at industry-related facts ("the organics industry is the fastest-growing segment of the American food system") and at specific points about you and your product ("we are located within one hour of the fastest-growing city in the state"). It can take place in several ways, depending on the product and the scale. You can investigate appropriate data, such as growth statistics from the census bureau or information on changing consumer interests from demographics services.

Talking to current consumers is another good way to learn about the market. For example, Kelly's Farm, which is already growing and selling produce and wants to expand into bottled products, might provide samples of proposed products to customers and then follow up with a survey. Often, you can find help with market research from area colleges that offer business degrees or from local and state government economic development offices.

A trip to the pumpkin patch is a fun family outing.

The last section, the financial plan, like the name says, talks dollars and cents. It provides a proposed budget and looks at key indicators, such as your company's *debt-to-equity ratio*, the relationship between dollars you've borrowed and dollars you've invested in your business. The more money you have invested in the business,

Visitors to your farm may enjoy the opportunity to meet some of your animals.

the easier it is to attract financing. Another key indicator is the *break-even analysis*, which calculates the volume of business necessary to break even; this is something that your creditors will want to know, but it's even more important for *you* to know.

For most aspiring businesspeople, working with a professional is the best way to develop a financial plan. Some states' cooperative extension services have farm business advisors who can help you. If yours does not, you may need to hire an accountant.

Business Structures

Sole proprietorship is a common form of business, in part because it is the easiest to set up and to terminate. One person (or a couple) owns and controls it, with funds for the business coming from the owner's personal funds, such as savings and/or investments, loans from lending agencies, and sometimes loans or monetary gifts from friends or family members. Personal assets such as land, homes, vehicles, bank accounts, and investments may be encumbered to pay damages resulting from lawsuits filed against the business or to pay claims by creditors for satisfaction of business taxes, loans, and contracts.

A *partnership* is a voluntary association of two or more people for the purpose of operating a business, with each person contributing money, property, labor, or skills to the business and each sharing in the business profits, losses, and liabilities. To protect others from using the partnership name, the partners must register the business name as a "doing business as" name in any state (or states) where the partnership operates. You can form partnerships with an oral or written agreement, but a written agreement is the better option.

There are three basic types of partnerships:

- **General partnership:** In this form of partnership, two or more people contribute assets to the partnership, and these general partners share the management, profits, and losses. The general partners manage the business and are personally liable for all partnership debts and liabilities and acts of any of the partners.
- **Limited liability partnership:** In this form of partnership, one or more partners limit their personal liability through written agreements between the partners. All partners may be general partners, with involvement in management decisions, but the liability of an individual partner is based on his or her personal areas of responsibility or the responsibility of someone under his or her direct supervision.
- **Limited partnership:** This is a way for the general partner(s) to acquire additional capital without giving up management control. In a limited partnership, there is at least one general partner and one or more limited partners. Limited partners take no active role in the management of the business, and their liability is limited to the extent of the money they contribute to the business.

Corporations have a legal and tax identity, separate from the owners or shareholders, and they are chartered by state governments. They must file articles of incorporation with the secretary of state where they receive their charter. Once formed, corporations have to file periodic reports with the secretary of state. Whereas the corporation is fully liable for all of its business obligations, individual shareholders are liable only to the extent of their investment. In practice, however, owners of small, closely held corporations (family members, for example) are often required to personally guarantee the debts of their corporation.

For income tax purposes, business owners must choose whether the corporation is a C or an S corporation. *C corporations* pay taxes on income, so their shareholders do not report any portion of corporate income or losses on their individual returns, but when income is passed on to the shareholders as salary or dividends, it is taxable income for them. A corporation with 75 or fewer employees may make a special election to be taxed as an *S corporation*. Although an S corporation must file a federal income tax return, it allocates income to the shareholders and taxes income at their personal state and federal rates. It must file an S election with the IRS on or before the fifteenth day of the third month of the first year that the corporation is chartered (try saying that three times very quickly!).

Limited liability companies (LLCs) provide the benefits of limited liability protection, operational flexibility, and pass-through taxation without the restrictions applied to S corporations and limited liability partnerships. The owners of an LLC are called members, and they have both economic and management rights (including voting rights) and may constitute the management of an LLC. As with corporations, most states that recognize LLCs require management to submit periodic reports to the secretary of state. Because LLC laws differ from state to state, take extreme care in drafting legal documents that meet the requirements of the state where you plan to do business. A generic form from the Internet may serve as a basic worksheet, but for your final draft, you'll want to know that your document is legally valid in your specific state. Contact a local accountant or lawyer for recommendations.

Selling Yourself

Developing your clientele is one of the toughest undertakings you'll make as an agripreneur, yet a loyal customer base—folks who come back again and again—are crucial to your business's success. These return customers are also your best advertisers. "Word of mouth" advertising grows a business better, and cheaper, than any other form of advertising.

Professional marketers use the terms *advertising* and *promotions*. Advertising is simply putting your business in the public's eye primarily through the use of some kind of media. Everything from

A farmers' market in Manhattan's Union Square attracts a crowd of city dwellers looking for farm-fresh produce.

Agritourism vacations offer visitors a chance to participate in the daily routine of a working farm.

business cards and billboards to social media to paid Internet ads falls under advertising. T-shirts or baseball caps with your farm's name on them are also a form of advertising, as is providing free samples.

Promotions are a specialized form of advertising, in which your business supports some community event or group. Sponsor a Little League team, and you are engaging in a promotion. Provide free or at-cost ground beef to the local Rotary for their annual community picnic or donate a certificate for farm produce to a silent auction at a local school, and you are doing a promotion.

One of the most successful, not to mention cheapest, advertising tools we have found for getting our business in front of new customers has been a computer-generated sign with our business name and phone number on "tear-offs" at the bottom. We placed these on bulletin boards in laundromats, in convenience stores, at the library, and in other readily accessible public places. Inevitably, each one of these signs garnered us at least one new customer.

Newspaper classifieds didn't net us many meat sales, although we did sell some breeding stock and calves using classifieds. However, some farmers have had success with marketing meat through newspaper advertising. Stephen and Kay Castner, of Kay's Home Farm Lean Meats in Cedarburg, Wisconsin, run a regular display ad in their local weekly paper, and they've also developed an online presence through their website (www.kaysfarm.com). The Castners raise purebred Galloway cattle on their farm, and contacts made through their website have generated breeding stock sales as far away as Kansas.

Many successful marketers encourage their clients to visit their farms. For example, some host an annual barbecue to which they invite all of their regular customers. As you'll read in the section on community supported agriculture (CSA) ventures, the Ocasions of Churn Creek Meadow Organic Farm have periodically hosted a harvest brunch featuring their farm's products. Others just extend a fairly open invitation. No matter what approach you take, the goal is always the same: create a proprietary interest in your operation among your customers.

Melvin and Carol Moon, of Moon Berry Farm and Puyallup Jam Factory in Puyallup, Washington, began offering a farm harvest tour at the urging of their extension agent. Now, the annual event is something they wouldn't consider forgoing. "At first," Carol says," we didn't want to do it, because harvest season is so busy, and we thought it would be too big of a responsibility to have crowds here. But one of the first people to come that first year said, 'We always wanted to visit here, and when we saw your tour advertised, it was like a personal invitation.'

"Once people have come on the tour, they feel more welcome—they almost take a sort of ownership interest in the farm—and they come back at other times to purchase from us, or they seek our products out in local stores."

Another strategy for developing a client base is doing promotions for civic groups. Offer to attend a meeting of the Rotary, a homemakers' club, or an environmental group. Prepare a Power Point presentation that describes your farm operation and the environmental benefits it brings. Have some sample products to give out after the presentation: meatballs in a slow cooker, hard-boiled eggs, or anything else you grow/raise and market. If your goal is to offer on-farm hunting or fishing, attend a sportsmen's club meeting or look into becoming a certified hunting safety instructor and then host those classes on site.

Free coverage is often available from local media because reporters (both television and print media) are always looking for stories. Prepare a media packet that will help them understand your operation and then invite a reporter to come to the farm and do a story. Sue Robbins of Pelindaba Farms on San Juan Island, Washington, has done just that, and her efforts have garnered both local and national coverage. She's even done a spot on the *Today Show* and has had coverage in *In Style* and *Better Homes and Gardens* magazines.

Sue offers some caution about going after media: "I have to tell you, most people have no clue how to deal with the press, so they do more harm than good. Your press releases and press packets need to be professional looking." She suggests that if you don't have experience in working with the media, you hire a consultant or join a local business association that can help you get some exposure with the media.

Once you've identified clients, keeping them becomes the next challenge. Find some way to keep in touch with regular customers. Hand out fliers and newsletters during presentations or e-mail them to your customer list. Tell people what's happening on your farm, tell stories about your animals, or let them know about food issues and policies that affect them and you—for instance, if a factory farm is trying to come into the area, keep customers apprised of developments.

Becky Weed, of Thirteen Mile Lamb and Wool in Montana, works hard on keeping in touch with customers, posting regular updates on her website (www.lambandwool.com). The updates are brief and written in a light and friendly fashion; they also include a few good-quality pictures of animals, scenery, or events from around the farm (including wildlife spottings). Mary and Albert Ocasion send out weekly e-mails to their CSA members, apprising them of what is fresh or special for that week. This TLC extends to printing weekly pages containing recipes, fun facts on a featured fruit or veggie, and more. These are tucked into members' boxes and relate to some aspect of the produce that the customers receive that week.

All correspondence with your customers should include information on ordering, product availability, new products, and any price changes, as well as how to get in touch with you.

Supplement your farm income by making and selling home-grown, homemade food products.

Spotlight on Community-Supported Agriculture (CSA)

The concept of CSA—community-supported-agriculture—was an idea whose time had come. During the 1960s, people were becoming increasingly concerned with the quality of foods available to them as consumers. A group of proactive neighborhood women in Japan took matters into their own hands, approaching farmers to establish a direct cooperative connection in which local growers were supported by local consumers. These women pioneered the Japanese equivalent of CSAs, the *teikei* (meaning "partnership" or "cooperation").

Meanwhile, in Europe, a similar movement was growing out of Robert Steiner's biodynamic farming, introduced earlier in the century. Steiner's concept was based on the idea that all living organisms have a dependent relationship. During the '60s in Switzerland and Holland, farmers cooperated to incorporate biodynamics into workable models similar to today's CSA, melding emotional, social, and economic components. Near Zurich, Switzerland, Jan VanderTuin cofounded Topanimbur, a CSA project and, in 1984, introduced the idea to Robyn Van En of Indian Line Farm in Massachusetts. Thus, the CSA concept came to America.

The underlying operatives of today's CSAs contain elements of the *teikei* model: cooperation between farmers and consumers; a focus on healthy, usually organic, products; and the principles of biodiversity in pursuit of that intangible symbiosis of interaction that mankind longs for—being part of each other and the land. According to the National Sustainable Agriculture Information Service, "A CSA consists of a community of individuals who pledge support to a farm operation so that the farmland becomes, either legally or spiritually, the community's farm, with the grower and consumers providing mutual support and sharing the risks and benefits of food production."

As part of the European model, involvement in a CSA project was seen as a way to help reconnect urban dwellers to the land and encourage a strong sense of community, including "…cooperation with a decided social justice goal to provide food security for disadvantaged groups." In some areas, CSAs are run as nonprofit organizations to complement food security programs. As such, they provide training and work for currently unemployed people, offer a venue for other local farms to be able to sell their products, and generate fresh produce for the food bank.

Another advantage to these nonprofit CSA operations is an inherent insurance against disruptions of food supply linked to urban areas, as well as a certain amount of farmland preservation through "green" practices. The ideals of *teikei* and biodiversity have intermingled, and, in some ways, CSAs, along with farmers' markets, agritourism, U-pick operations, and farm stands, are part of this century's rediscovery of a more agrarian society despite urban sprawl and its issues. As

If you grow enough produce in variety and quantity, you might offer your goods for sale through a CSA subscription program.

members, city dwellers can participate on several levels in the growing, choosing, and sometimes picking of their own food. This food, sustainably and responsibly grown without harmful chemicals or other unhealthy practices, is helping address some of what we lost with the nation's shift away from farm-based living. According to a research paper published by the University of California-Davis, CSA addresses problems at the nexus of agriculture, environment, and society. These issues include a decrease in the percentage of the food dollar that goes to the farmers who do the work, financial obstacles to new farm start-ups, large-scale implications of food-borne illnesses, and degradation and/or depletion of resources and environment.

How CSAs Work

There is no one "right" way to run a CSA, but there are some basic patterns, falling roughly into three categories:

- The farmer owns the land and machinery and does all of the work, utilizing capital from members/shareholders.
- The farmer owns the land and equipment, and the members/shareholders provide capital and part of the labor.
- There is no farmer. The members/shareholders own the land and equipment and are responsible for growing their own produce.

Each of these categories gives rise to variations and combinations as people figure out what they need and how best to implement methods to meet those needs. As with any new venture, if you're considering launching a CSA, be sure to do your homework. There will be legal requirements. For example, if marketing your produce as organic, you must comply with the requirements of producing organic food. If you are considering a CSA operation, there are multiple resources available, including some at the end of this book.

Setting It Up
Membership/Shares Model

In what is referred to as the "membership/shares" model, consumers pay up-front for part or all of the farm products that they will receive from the farmer throughout the season. This can range from approximately $150 to $600 (or more) for a season, depending on the size of the share and the length of the harvest. This money, committed in advance, enables the farmers to purchase necessary seeds, repair or replace equipment, and deal with other farm-related expenses, thus empowering them to successfully produce the food for which their shareholders have prepaid.

A "share" usually includes seven to ten choices of fresh vegetables and/or fruits, enough to supply two or three people with a week of produce. By prepaying a certain amount, members have "credit" with the farmer and are able to pick and choose how they want to apply that credit within the offered structure of the particular CSA. In addition to produce, some CSAs offer opportunities to own a share in a farm animal. With this up-front financial outlay, CSA members provide financial support to the farm operation and farmer's salary; in return, members receive not only shares in the farm's products but also the satisfaction of experiencing the land and a more agrarian lifestyle.

Obviously, this is a win-win situation both for farmers and for their CSA members/shareholders. The latter are reconnected with the magic of the land and are assured top-quality, locally grown products—often including free-range eggs and/or organically raised, grass-fed beef and other meats, including chicken and turkey—while, according to the USDA Natural Agriculture Library, "Growers receive better prices for their crops, gain some financial security, and are relieved of much of the burden of marketing by selling directly to community members." Many CSAs offer payment

Covered Bridge Produce

Joseph Griffin runs Covered Bridge Produce in Oley, Pennsylvania. This CSA is a six-figure success story with 400 clients and almost a dozen employees during the growing season. With seven acres under intensive cultivation on a twenty-six-acre farm, Joseph oversees the growing, harvesting, packaging, and delivery of more than 130 varieties of vegetables, ranging from arugula to zucchini. He buys fruit, flowers, and sweet corn from other local growers, freeing him and his staff to concentrate on what they do best.

Joseph was a professor and college administrator who "sank to the level of a fundraiser" when he got tired of it all and dropped out of the system, buying his farm in 1996. He played around at first—gardening and selling a little produce at farmers' markets. But, in 2000, as he watched his retirement funds draining away, he decided that he either needed to have the farm make some real money or find something else that would.

He began looking for a clientele that would support his business, and he set upon a modified CSA model as the best way to market to them. "We looked for a kind of a 'fuzzy' clientele that is really concerned about its health and that has disposable income. These are people that like to cook, and that are really interested in local, organic food."

Joseph's clients go to the website each week, and they get to choose up to ten items from a list that may include upward of seventy-five products. The website records their choices in a database and then prints out sheets for harvesters and packagers, making for accurate and speedy work. This system saves Joseph about 40 percent over the cost of harvesting and packaging before the system was in place. The system also allows Joseph to plan better for the next season based on the buying decisions that his clients made the previous year.

plans and have been approved to accept food-assistance vouchers, which enables people who wouldn't otherwise be able to purchase such high-quality food to become members. Other farms may allow a sliding scale or offer scholarship shares.

Box Model

The "box" model, by far the most widely used way of organizing a CSA, is similar in many ways to the membership/shares model, varying primarily in that members do not purchase shares in the farm or animals. The focus does not, then, have as much of the "our farm" dynamics. Rather,

The contents of a CSA box vary from week to week or month to month based on growing seasons and harvests.

members prepurchase produce and farm products by the box. Individual CSAs may permit weekly, biweekly, or monthly purchase, whereas others stipulate larger payments less frequently, with the median time period being prepaying every eight weeks. This particular aspect varies little from the member/shareholder model because, with prepayment, farmers still have the financial resources to grow the produce and maintain their equipment, but they do not get the full season's up-front guaranteed payment.

Why go this way if the membership/share model is more secure? It's not that simple, actually, because many consumers do not have the ready capital to plunk down several hundred dollars at the beginning of the season; therefore, they won't/can't join. With the box method, although there is more financial uncertainty for the farmer, getting and keeping customers is much more

feasible, which is why some CSAs opt for weekly or biweekly payments, knowing that it is a lot easier for customers to pay $25 or $50 dollars at a time than several hundred.

Within the box model are a few subcategories: single farm box, collaborative box, and farm-linked aggregator box. A single farm box is what the name implies—a box generated from a single farm. However, many single-farm operations will occasionally offer their members products produced by other farms. Fruits in season from local orchards or vineyards or eggs from a neighboring organic chicken farmer

CSA as we know it today can be traced back to the Zurich region of Switzerland.

Getting Started

As an overview, the University of Florida's Institute of Food and Agricultural Sciences Extension offers helpful steps to launching your own CSA:

1. Farmers or groups of nonfarmers initiate the process by issuing a call to form a CSA.

2. Prospective farmer(s) and sharers hold an exploratory meeting.

3. At the exploratory meeting or a follow-up meeting, farmer(s) and sharers come to an agreement on the group's values and goals.

4. Organize the core group, which is a council of growers and members who will work together to run the CSA. This group determines the duties, activities, and functions of the group, and their responsibilities may include deciding which crops to grow, what fees will be paid and by whom, and when payments are due. They will also select farmers and/or land to be used and determine the distribution process.

5. The core group recruits members for the initial season.

6. Members make commitments to the new CSA, including purchasing monthly shares and delegating tasks (e.g., sorting and packing shares, coordinating delivery or pickup).

7. The core group establishes the legal status of the CSA, although they may defer decisions on legal structure for the first couple of seasons. Some CSAs separate the CSA from the ownership of the land, which may be held as a corporation, sole proprietorship, or a partnership of varying definitions. At this time, there is not a set structure in the law for buying clubs or food co-ops, so legal details will vary from state to state; a lawyer's advice can help.

8. The core group decides on capitalization of the farm; this is where the model of prepaying for shares is of real assistance. In the short term, the CSA may start with a minimum of equipment, but the group will need to discuss and make decisions on equipment purchases and maintenance for the long term. For example, if you as a farmer are operating as owner, with a membership of subscribers, you will have a different model than that in which there is no farmer, and the members are responsible for the land, machinery, and workload.

Author's note: Most CSAs tend to carry standard liability insurance as part of their farm insurance package. Purchased separately, it can be much pricier. If you plan on operating a U-pick CSA, you may need additional liability coverage. Fees can be lower if you are an organic farm or if pickers don't use hazards such as ladders, horses, or machines.

are common offerings. A collaborative box CSA is what its name implies: several farms working together to manage the CSA and market their products. Sometimes the organizers of collaborative CSAs are independent of the farms themselves. The *aggregator model* is similar to the single-farm-box type of CSA in that it is linked to a single farm. The difference is that these CSAs consistently offer products purchased from other farms and/or wholesale markets.

Albert and Mary Ocasion of Churn Creek Meadow Organic Farm, Redding, California, are an example of a farmer-owned subscription-based CSA run according to a combination of single-farm and aggregate-box method. The Ocasions launched their CSA in 2008. Three years prior, Mary's love of gardening resulted in excess produce, which she began selling at local stores and the farmers' market. "Our business is the farm and selling produce at the farmers' market, plus the CSA, plus wholesale," says Mary.

In addition to their weekly subscribers, the Ocasions wholesale to a small grocery chain (Holiday Market) that likes to carry local organic produce. This gives Churn Creek Meadow a reliable market for their extra garden bounty. "There were some years that we had so much that it just worked out great for us. One year, we brought them a truckload of watermelons. Another year, it was honeydew.

They're flexible. Our normal [products] for them are tomatoes in the little green mesh bags, the mixed heirlooms, lemon cucumbers, and eggplant and basil."

Every week, Mary cuts each arm of every plant on her 200-foot row of basil to keep them from bolting (going to seed). "I pick it, put it right in water, and then bundle it and bring it to the market—farmers' market and Holiday—in water so that they stay fresh. People appreciate that." Some box CSAs may offer boxes of different sizes and content, with prices corresponding. Others base their boxes on what is currently in season. Therefore, if zucchini is in and strawberries are out, alas! If strawberry shortcake was on your menu, you'll have to get the fruit elsewhere or settle for zucchini bread.

At Churn Creek Meadow Organic Farm, the goal is to deliver farm-fresh organic produce to customers year-round. However, the climate in their part of Northern California means that although they stretch the growing season to the utmost, in reality, year-round growing is not always possible. Mary says,

A typical CSA share includes a variety of fruits, vegetables, and even herbs.

"Our herbs go into the winter, and our winter squash goes until frost, but our business is year-round, so we buy from other farms." Even in the summer, they're buying from other farms in order to offer items they don't grow themselves, such as orchard fruits and some root crops. Mary gestures to a stack of boxes just delivered. "Customer really love the organic bananas, so we have some. But other than bananas, we try for California, local, as close to home as possible, and for about half the year, that's pretty doable. We try to get organic apples from California, but when there are none, we bring a shipment in from Washington. We'll go for the California pears and apples until the supplies are depleted from the growers we know, and then we'll switch to Washington. Our potatoes come from southern Oregon." The Ocasions strive to obtain the least traveled produce as possible per season. Mary pampers their customers. "That keeps them with us."

Churn Creek Meadow CSA has about 150 members who pay by the week, although some opt to pay ahead. Customers make payments by check or via credit card. As part of their very individualized CSA, Mary and Albert supply customers every Sunday with what they call the "extras list," apprising their members of added options for the coming week. This is a rather labor-intensive practice, but it has paid off in customer satisfaction. Mary shares, "The first year we didn't do that, since most CSAs don't, and we lost a lot of customers because they want certain things. They might be allergic to something we offer, or they might grow it themselves in their own garden. Some people grow tomatoes and nothing else. Some have a full-blown summer garden but still want other things, so they might order a fruit box. We've got one person who has a fruit orchard, and he gets just vegetables. So we try to supply what they want."

Some CSAs take a community-gardening approach in which the members grow their own produce and share with one another.

All across the country, CSA owners are adding personal touches to their operations and thereby creating loyal customers. Some, like the Ocasions, include recipes in their weekly boxes to give buyers ideas on how to use produce that may be new to them. In addition to continually updating website information by adding photos shot on location, information on what they're currently planting, health trivia, featured products, and more, growers work hard to make sure that their CSA customers receive their money's worth and come away feeling a connection to the farm and its bounty.

Customizing Your CSA

Although many CSAs offer only produce, as previously mentioned, some have branched out to offering meats (beef, poultry), eggs, raw milk (make sure you comply with legal regulations regarding raw milk), and dairy products such as butter and cheese. These can be regular offerings or made available to customers as they come available, as is the case at Churn Creek Meadow. Mary and Albert offer grass-fed, antibiotic-free beef halves and quarters as options for CSA customers. "We put out that blurb, and I have a waiting list of people that want halves and quarters." If they don't presell all of the meat, they can have the steers butchered by the USDA butcher, freeze the meat, and then sell the meat at the farmers' market. The Ocasions also sell eggs at their CSA. "Our hens are free range, but with the cost of organic feed, they aren't certified organic. However, they eat all of our leftover [organic] produce, which really adds to the sustainability of the operation."

Many CSAs offer farm tours and activities, where visitors can come out and experience the gardens, the herbs, the animals, the pumpkin patches or corn mazes—whatever the CSAs feature. This looks a lot like agritourism, but with the added element of farm-grown products. These additions can be great ways to spread the word about your CSA. In their own variation, for several years, Mary and Albert hosted a farm day on Labor Day, a ticketed event featuring food from their farm, along with outsourced organic meat, in a catered brunch. Around Thanksgiving, they also put together a special box that includes sweet potatoes, beans, and other traditional offerings. This is another area in which CSAs mix and match to come up with something that works well for their unique operation.

Some CSAs specify pickup only, which means that members come to the farm to pick up their boxes. Others may offer a delivery service to designated dropoff sites in addition to allowing shareholders to pick up at the farm. Churn Creek Meadow CSA is unique in that it offers a complete delivery service within 15 miles of the farm. "Having people come here to pick up produce wasn't an option when I was working a full-time job in addition to the CSA," says Mary. Now that she devotes her time to their CSA, customer pickup on the farm is more feasible, but Albert prefers to deliver to homes and offices. "Like the milk man," Mary laughs. "We get more business that way, because delivery really enables customers to take part in a CSA, where they might not otherwise be able to make the dropoff schedules. He'll leave the box of produce on the front porch and pick up the check under the mat. Or sometimes customers will leave an ice chest out with a cold pack inside and the check taped under the lid."

As mentioned before, payment schedules also vary as CSA owners seek to balance the bottom line and still offer flexibility to their members. Concerning their payment method, Mary says, "With us, members just sign up on the web to receive their box [that] week if they want to, not six or eight weeks like some that pay ahead in order to be sure of the customers will show up. We can be more flexible. If by noon on Monday you notify us and say, 'Hey, I need to skip this week,' it works all right because we have the stores and farmers' market to siphon off the extras." She adds that, in the summer, the farmers' market makes up about half of their farm sales, and they discontinue selling there after frost reduces their farm's produce offerings.

"For our CSA customers, we list on the website how many of each thing they will receive in each type of box. We're pretty comparable, price per pound, with what they would pay at the farmers' market or the store. Locally grown, very fresh. People say it lasts longer than store-bought produce in their refrigerator, and that's because it was picked more recently." Mary also orders from the other farms at the very last minute so that they can have a good idea of what they really need and pass that last-minute freshness on to members.

As an example of customizing products, the Ocasions offer six different boxes—among them an all-fruit box and an all-raw eating box (containing a mix of fruits and vegetables, but nothing

CSA shares are not limited to just produce. Depending on the farm, members might receive eggs and/or dairy products, too.

Vala's Pumpkin Patch

Agritainment: A marriage of agriculture and entertainment into a fun experience for the whole family.

Jan and Tim Vala of Vala's Pumpkin Patch in Gretna, Nebraska, are pioneers of agritainment. Back in the early 1980s, at a time when farmers did little direct marketing, pick-your-own operators were exploring new territory. Tim, burning with a desire to have a profitable farming operation, thought this new approach to agriculture held promise for his family's future. He went around the country to visit with farmers who were experimenting with pick-your-own operations, and then he went in search of a farm suitable for a pick-your-own strawberry operation. He found one—less than thirty miles from Omaha, Nebraska, and just a little farther from Lincoln.

With twenty acres of strawberries, the Valas were barely breaking even. Then, one year, Tim planted a field of pumpkins, too, as a way to diversify his operation. He sold the pumpkins out of the back of his pickup truck, down by the highway. People seemed to like that, so he planted more pumpkins the next year. But this time, he had too many to sell by the road, so he used the pick-your-own approach and had buyers come to the farm. He would take them on a hayrack to the field, and the weather was usually pretty nice.

Today, the Valas' thirty-five-day pick-your-own-pumpkin season attracts as many as 12,000 visitors on a Saturday afternoon. With a small staff of year-round helpers and more than 300 seasonal employees who run a dozen food booths, act the parts of ghosts and goblins, and help with pumpkin sales, they have created a very successful business model. They charge admission by the day or by the season and sell fifty-five acres worth of pumpkins to visitors from far and wide.

Jan tells me that marketing has been crucial to their success. "There is a lot of advertising you have to do, though once you have a good reputation, it's a lot easier. In the early years, we would do newspaper ads, radio and cable television spots, and billboards. We still have a brochure that we print every year, and I have a company that distributes them to hotels, restaurants, gas stations, and along the interstates. Those are geared to getting new people to come to the farm."

The Valas have also worked to develop corporate and business supporters, who buy passes to give away to their customers or employees. Jan provides an example of an Omaha car dealership that purchases passes in bulk and then gives them away to potential buyers who test-drive a car. She emphasizes that these business connections provide stability and early-season cash flow that helps keep their business strong.

you have to cook, like potatoes). They offer a "juicing box" and two different sizes of mixed boxes. Customers can also purchase gift certificates or a gift box.

How much customer service do you want to provide? If you just want a midweek market and you feel like you can charge people for six, eight, or twelve weeks of produce ahead of time, you might find it more convenient because you know that they're going to come pick up their boxes. They're not going to cancel on you on Monday, as they might with Churn Creek Meadow Organic Farm.

As with every business, CSAs have a few untidy ends that farmers answer creatively. Adaptability is the key word here. One of these areas is containers. If the CSA furnishes boxes, they can run as high as $3 or more apiece, and not all customers return them. A box deposit may be a way to cut down on this expense, but presents its own complications. A bring-your-own-box or bag isn't the best solution, since the produce must be picked and packed at certain times, and the customer may forget to make sure that his or her box is at your farm at those times.

Surviving in the off season is also often a concern. Cash flow drops as harvest ends. This time can be productive in other ways, though. In addition to setting up the land for the next growing season, there may be additional time to send out flyers, update websites and advertising, and plan for next year. Those who follow collaborative or aggregate models may choose to keep the CSA going in the off season through selective importing from other regions. Mary says, "We continue on with the CSA. We just order more importing at the same price. We make a lot less money, but there's less labor, so there's somewhat of a trade-off at that point." Interestingly, she also noted that their boxes-per-week number rises in the winter. "People are more inclined to be home, and the farmers' market is closed. Also, in January, we see that number go up because of New Year's resolutions. People resolve to be healthier and cook and eat more at home. The numbers then slope off to more normal levels by May or June as members leave for vacation or their own gardens start to produce. Some regulars put their memberships on hold until the fall."

There's always plenty to do on the farm.

The Labor

In case you don't already know it, farmers are some of the hardest working people you'll ever run across, and CSA farmers are no exception. Readying the soil, planting, weeding, cultivating, picking packaging, selling—these all come into play in running a CSA. "It's good exercise, and that's all part of it for me," says Ocasion. "I like the exercise. I think that's part of why I like to farm."

Depending on how much land you'll be planting, you may be able to get along without a lot of initial outlay for machinery. For under an acre, a good quality tiller may suffice. Beyond that, a mid-sized tractor may be on your wish list, along with a PTO-driven rototiller and other implements as needed (see Chapter 3 for info on tractors). CSA members may be a good resource for farm labor, even if your CSA is not organized in that fashion. On an individual basis, some members may desire to be more hands-on, involved in the nitty-gritty of growing things, and may want to trade labor for shares or part shares. One single mom who loved working at a local CSA eventually became a valued employee there.

Although there's no doubt that operating a CSA is work and a lot of it, for many who have chosen that route, it is very rewarding. As Mary Ocasion puts it, "People are very appreciative. We get the nicest emails from these families. It's rewarding because you're providing them a way to be healthier. Not only is it a healthy lifestyle change for a lot of people, we can make a living doing what we love while providing them with something they appreciate."

Resources

Books

Ken and I have shelves and shelves of farming books that we have collected with zeal over the years; our oldest dates to the 1860s. Some of my favorites are:

Biodiversity Heritage Library. Download hundreds of vintage farming and livestock books in free PDF format. www.biodiversitylibrary.org

Bubel, Nancy. *The New Seed-Starters Handbook*. Emmaus, PA: Rodale Press, 1988.

Caldwell, Gianaclis. *The Small-Scale Cheese Business*. White River Junction, VT: Chelsea Green, 2014.

—. The *Small-Scale Dairy*. White River Junction, VT: Chelsea Green, 2014.

Coleman, Elliot. *Four Season Harvest*. White River Junction, VT: Chelsea Green, 1999.

—. *The New Organic Grower*. White River Junction, VT: Chelsea Green, 1995.

Damerow, Gail. *Fences for Pasture and Garden*. North Adams, MA: Storey Books, 1992.

—. *Storey's Guide to Raising Chickens*. North Adams, MA: Storey Books, 2000.

Ekarius, Carol. *Small-Scale Livestock Farming: A Grass-Based Approach to Health, Sustainability, and Profit*. North Adams, MA: Storey Books, 1999.

Kowalchik, Claire and William H. Hylton, eds. *Rodale's Illustrated Encyclopedia of Herbs*. Emmaus, PA: Rodale Press, 1998.

Logsdon, Gene. *The Contrary Farmer*. White River Junction, VT: Chelsea Green, 1995.
Author's note: Gene Logsdon has inspired us and helped us learn how to do many of the things that we've done. For example, his

Organic Orcharding book is a must-read for those interested in seriously pursuing an orchard, and his *Small-Scale Grain Raising* is great for anyone who wants to grow grain crops, such as wheat or corn. Read any and all books by Gene Logsdon that you can get your hands on (many are now out of print, but you can often purchase them used or find them in libraries).

Mettler, Dr. John J. *Basic Butchering of Livestock and Game*. North Adams, MA: A Garden Way Publication, Storey Books, 1986.

Poisson, Leandra, and Gretchen Vogel. *Solar Gardening: Growing Vegetables the American Intensive Way*. White River Junction, VT: Chelsea Green, 1994.

Salatin, Joel. *Family Friendly Farming*. White River Junction, VT: Chelsea Green, 2001.

—. *Pastured Poultry Profits*. White River Junction, VT: Chelsea Green, 1996.

—. *Salad Bar Beef*. White River Junction, VT: Chelsea Green, 1996.

—. *You Can Farm*. White River Junction, VT: Chelsea Green, 1998.

Schwenke, Karl. *Successful Small-Scale Farming*. North Adams, MA: Storey Books,1991.

Simmons, Paula, and Carol Ekarius. *Storey's Guide to Raising Sheep*. North Adams, MA: Storey Books, 2000.

Spaulding, C. E. *Veterinary Guide for Animal Owners*. North Adams, MA: Storey Books, 1996.

Periodicals

Acres USA
www.acresusa.com/magazine
(800) 355-5313

The Caretaker Gazette
(206) 462-1818
www.caretaker.org

Countryside
(800) 551-5691
www.countrysidemag.com

Growing for Market
(800) 307-8949
www.growingformarket.com

Hobby Farms
Contact form under "Contact Us" on the website
www.hobbyfarms.com

Small Farm Today
(800) 633-2535
www.smallfarmtoday.com

Small Farmer's Journal
(800) 876-2893
http://smallfarmersjournal.com

The Stockman Grass Farmer
(800) 748-9808
www.stockmangrassfarmer.com

Organizations and Government Resources

Agricultural Marketing Resource Center (AgMRC)
This is a comprehensive resource on funding and growing a value-added agriculture business.
(866) 277-5567
www.agmrc.org

Alternative Farming Systems Information Center (AFSIC)
AFSIC is an arm of the USDA that focuses on sustainable agricultural practices.
(301) 504-6559
http://afsic.nal.usda.gov

Alternative Technology Transfer for Rural America (ATTRA)
ATTRA is funded by the USDA and is managed by the National Center for Appropriate Technology (NCAT). It provides information and technical assistance to farmers, ranchers, extension agents, educators, and others involved in sustainable agriculture in the United States. It is one of the best resources I know of, with outstanding information available online or mailed to you free of charge.
(800) 346-9140
https://attra.ncat.org

Center for Rural Affairs
This organization lobbies for policies that improve the lives of family farmers, small business owners, and rural citizens.
(402) 687-2100
www.cfra.org

Cooperative Extension System
The Cooperative Extension System is a nationwide educational network comprising an office for every state as well as a network of local and regional offices. Offices are staffed by one or more experts who provide useful, practical, and research-based information to agricultural producers, small business owners, youth, consumers, and others in rural areas and communities of all sizes.
(202) 720-4423
www.csrees.usda.gov/extension

Cornell Small Farms Program
Cornell University's organization functions in support of small farms, sustainable practices, research, and extension programs focused on rural communities.
(607) 255-9227
http://smallfarms.cornell.edu

Federal Emergency Management Agency (FEMA)
The US government's national agency for disaster preparedness, response, and recovery.
(202) 646-2500
www.fema.gov

Kerr Center for Sustainable Agriculture
A nonprofit educational foundation offering events, research, publications, and assistance to rural communities in its home state of Ohio, as well as to farmers and rural residents across the country.
(918) 647-9123
www.kerrcenter.com

The Livestock Conservancy (formerly the American Livestock Breeds Conservancy)
The Livestock Conservancy is dedicated to preserving heritage breeds of livestock.
(919) 542-5704
www.livestockconservancy.org

National Environmental Services Center
A West Virginia University-based service with the goal of helping rural residents with issue related to drinking water and wastewater management.
(800) 624-8301
www.nesc.wvu.edu

National Renewable Energy Laboratory (NREL)
NREL is the nation's leading laboratory for renewable energy research and development.
(800) 363-3732
www.nrel.gov

Natural Resources Conservation Service (NRCS)
This branch of the USDA provides farmers and ranchers with financial and technical assistance in making conservation-oriented improvements to their land and implementing government conservation programs. They have many resources available; for example, "Ponds—Planning, Design, Construction" (Agriculture Handbook 590) is an excellent guide to choosing your site, designing your pond, and incorporating spillways and dams. Call or look on the website to find your local office.
(303) 275-3000
www.nrcs.usda.gov

Organic Crop Improvement Association (OCIA)
OCIA provides research, education, and certification services to thousands of organic food growers, processors, and handlers.
(402) 477-2323
www.ocia.org

Sustainable Agriculture Research and Education (SARE)
SARE is an initiative funded by the USDA that sponsors competitive grants for sustainable agriculture research and education. Check out the CSA directory on the SARE website along with links to the websites for SARE's central, southern, western, and northeastern divisions.
www.sare.org

United States Geological Survey (USGS)
A Virginia-based organization with offices and agents in all regions of the country with the goal of providing scientific information on the environment, ecosystems, climate, natural resources, and related topics.
(888) 275-8747
www.usgs.gov

Commercial Providers

There are hundreds of excellent commercial providers for tools, equipment, seeds, and other farming resources. Here is a selection of my favorites:

Jarden Home Brands
The company behind Ball and Kerr canning jars as well as other food-preservation appliances, tools, and products.
(800) 240-3340
www.homecanning.com

Johnny's Selected Seeds
Johnny's is a great resource for seed for both gardeners and field-crop producers. The company carries vegetable and flower seeds; culinary, medicinal, and aromatic herb seeds; gardening tools; garden supplies; and home garden accessories. Johnny's thoroughly tests all of its seeds and accessories at the company's certified organic farm in Albion, Maine, and backs up its products with a 100-percent satisfaction guarantee.
(877) 564-6697
www.johnnyseeds.com

Lehman's
Lehman's—a company that caters to the Amish community—is the source for old-fashioned, hard-to-find items such as pickle kegs, grain mills, wooden barrels, hand water pumps, wood cook stoves, heating stoves, canning supplies, and much more.
(888) 438-5346
www.lehmans.com

NASCO

NASCO's catalog is like the Sears catalog for farming.
(800) 558-9595
www.enasco.com

Premier Fencing

Premier makes high-quality fencing materials that are great for all applications.
(800) 282-6631
www.premier1supplies.com

Seeds of Change

This is an excellent source for garden and farm seeds, as well as for tools, equipment, and other goods.
(888) 762-7333
www.seedsofchange.com

World Wide Opportunities on Organic Farms

Organization that pairs volunteer workers with organic farmers, who host the workers and provide room and board in exchange for daily help on the farm.
(415) 621-3276
www.wwoof.net

Online Resources

Agritourism

Agricultural Marketing Center: Agritourism
www.agmrc.org/commodities__products/agritourism

ATTRA: Entertainment Farming and Agritourism
https://attra.ncat.org/attra-pub/summaries/summary.php?pub=264

North Carolina Department of Agriculture and Consumer Services: So You Want to Start an Agritourism Farm?
www.ncagr.gov/markets/agritourism/documents/StartingAgritourismBusiness.pdf

University of Vermont Agritourism Resources
www.uvm.edu/tourismresearch/?Page=agritourism.html

Beginning Farmers

Beginning Farmers Website
www.beginningfarmers.org

Iowa State University Beginning Farmer Center
www.extension.iastate.edu/bfc

Community-Supported Agriculture

Community-Supported Agriculture Resource Guide for Farmers
http://growingsmallfarms.ces.ncsu.edu/growingsmallfarms-csaguide

University of Tennessee Institute of Agriculture: A Farmer's Guide to Marketing through Community-Supported Agriculture (CSA)
https://utextension.tennessee.edu/publications/documents/pb1797.pdf

Dairying

Cornell Small Farms Program: Small Dairy
http://smallfarms.cornell.edu/resources/small-dairy

Fias Co Farm: Home Dairying, Milk, and Cheesemaking
http://fiascofarm.com/dairy

Small Dairy Website
www.smalldairy.info

Gardening

Cornell University: Cornell Garden-Based Learning
www.gardening.cornell.edu

National Garden Bureau
http://ngb.org

University of California Division of Agriculture and Natural Resources: Gardening
http://ucanr.edu/Gardening

University of Wyoming Extension: Master Gardener
www.uwyo.edu/mastergardener

Livestock

University of Minnesota Extension: Alternative and Small-Scale Livestock Systems
www.extension.umn.edu/food/small-farms/livestock

North Carolina State University Cooperative Extension: Small-Scale Livestock
http://growingsmallfarms.ces.ncsu.edu/growingsmallfarms-animallinks

University of Florida/Florida A&M University: Livestock
http://smallfarms.ifas.ufl.edu/livestock_and_forages/livestock_index.html

Livestock Predation

Internet Center for Wildlife Damage Management: Livestock and Animal Predation Identification
http://icwdm.org/inspection/livestock.aspx

Maryland Small Ruminant Page: Predators and Predation
www.sheepandgoat.com/predator.html

Poisonous Plants

Cornell University: Plants Poisonous to Livestock and Other Animals Database
www.ansci.cornell.edu/plants

Poultry

North Carolina State Cooperative Extension: Backyard Flocks and Eggs
http://poultry.ces.ncsu.edu/backyard-flocks-eggs

Purdue University Agriculture: Poultry Page
https://ag.purdue.edu/ansc/poultry/Pages/Resources.aspx

Preserving Food

National Center for Home Food Preservation
http://nchfp.uga.edu

Property

University of Wisconsin Extension: Country Acres—A Guide to Buying and Managing Rural Property
http://learningstore.uwex.edu/assets/pdfs/G3309.pdf

Oregon State University: Top Ten Things I Learned Buying a Small Farm
http://smallfarms.oregonstate.edu/sfn/fall06topten

Oklahoma Cooperative Extension Service: Moving to the Country?
http://pods.dasnr.okstate.edu/docushare/dsweb/Get/Document-2530/AGEC-965web.pdf

Wind Power

National Renewable Energy Laboratory—Wind Energy for Farmers and Ranchers
www.nrel.gov/learning/re_wind.html

US Department of Energy: Wind for Homeowners, Farmers, and Businesses
http://apps2.eere.energy.gov/wind/windexchange/small_wind.asp

Index

Note: Page numbers in **bold** typeface indicate a photograph.

vegetables, miniature varieties, 131–133, **131–132**
  . *See also* fruits and vegetable harvest
vehicles, 54–55, **54**
  . *See also* tractors
ventilation, in barns, 38
veterinarians, large-animal, 52
A Veterinary Guide for Animal Owners (Spaulding), 147
victory gardens, 126–127, **126**
viral infection, 159–161
Vitaceae plant family, 100
volunteer opportunities, 18

W

walking tractors, 93
warm stratification, 104
warranty deeds, 29–30
water, condition and safety, 25
water conservation, 27
water cycle, 85
water requirements
  barn location, 41
  chickens, 184, **184**
  disaster preparedness, 81

farm animals, 155, **155**, 157
farm location, 23–26, **23**
flower beds, 118, **118**
winter maintenance, 48–49
water troughs, 48
water-bath canning process, 229–230, **231–232**
waterers
  automatic, 158, 183
  chickens, 178, 183, **184**, 185, 191–193
  gravity-fed, 183
  heated, 179
  pigs, **155**
weather exposure, 118
Weed, Becky, 70, 277
weed control
  biological techniques, 67–68, **67**
  burning, 66–67
  herbicides, 68
  mechanical techniques, 64–66, **64**
  mowing, 66–67, **66**
  overview, 63–64, **65**
weeds, types of, 65, 68
wellhouses, 45

western barn style, 35–36, **36**
wheelbarrows, 55
wheezing, 160
White, E. B., 140
wide-row gardening, 93
Wilson administration, 51
wind generators, 28, **29**
windrowed hay, 111
winter annual weeds, 65
winter chores, 48–49, **48**
wiring, in barns, 38
wool, poor-quality, 160
worker bees, 204–207, **205**
workspaces, in barns, 39
World Wide Opportunities on Organic Farms (WWOOF), 18

Y

Yankee barn style, 33
yogurt, 254–255, **254**
yolkless eggs, 186

Z

Zody, Michael, 169
zoning laws, 25, 32, 45, 173, 199

Photo Credits

About the Authors

Carol Ekarius is a leading writer on sustainable agriculture and hobby farming, emphasizing practices that are good for the environment and the farmer's bottom line. She is a regular contributor to *Hobby Farms* magazine. She and her husband Ken have farmed for more than twenty years (nine years as commercial farmers), raising cattle, sheep, pigs, horses, mules, donkeys, poultry, and fowl. They reside in Hartsel, Colorado, near Denver.

Freelance writer **Leslie J. Wyatt** writes for both children and adults and has more than 200 articles and stories in various magazines, websites, and anthologies. A regular contributor to *Hobby Farms* and hobbyfarms.com, she also has three middle-grade novels in print and two more in revision. She enjoys spending time with her husband of 32 years and her two teenage daughters; riding her Arab/Quarterhorse, Honor; and exploring various creative outlets.